The Dimensional Structure
of Consciousness

The Dimensional Structure of Consciousness

A Physical Basis for Immaterialism

Samuel Avery

compari

✛

lexington/kentucky

1995
COMPARI
308 Madison Place
Lexington, Kentucky
40508-2516

Copyright © 1995 by Samuel Avery

Printed in the United States of America

Cover Graphic: Joshua Nespodzany

Library of Congress Cataloging-in-Publication Data

Avery, Samuel, 1949-
    The dimensional structure of consciousness : a physical basis immaterialism / Samuel Avery.
        p.   cm.
    ISBN 0-9646291-0-0 (pbk.)
    1. Physics—Methodology. 2. Physics—Philosophy.
3. Consciousness. 4. Immaterialism (Philosophy) I. Title.
QC6.A88   1995
113—dc20   95-41878
                                                           CIP

CONTENTS

I
INTRODUCTION

S cience is the great power of western civilization. The power of science is its ability to establish new truths without reference to fundamentals, to progress beyond prejudice and opinion, and to open the physical world to human manipulation. The great weakness of science is its inability to include all of experience. Thought, imagination, and experience itself cannot be the proper subjects of scientific study, and are thereby excluded from the world presented by science. This is the great weakness of western civilization.

It is my belief that the world of science is not the world itself, but a special structure of conscious experience within a larger context. I have come to this conclusion because of what science itself, particularly modern physics, has discovered about nature. We have learned through quantum mechanics, for example, that extremely small pieces of matter are not "in" space and time in the normal sense, and that there are "particles" which have no mass, electrical charge, intrinsic size or shape. Through relativity theory we have learned that a moving object is prevented by a property of empty space from traveling faster than light, and that as it approaches the speed of light, its time slows and mass increases. The possibilities of absolute time and absolute measure no longer exist. Science can explain this only with great violence to the concept of an external, material world that we are "in." I hope to show in this book that the findings of science make better sense if this concept is eliminated entirely.

This is no more nor less than good scientific procedure. If findings do not fit theory, it is toward assumptions underlying theory that we should look. Matter is an assumption underlying physics. It serves to explain experience, but is never itself experienced directly. Everyone sees, hears, and touches physical objects, but no one sees, hears, or touches matter in any absolute sense. We experience visual and tactile perceptions that suggest a material substance existing independently of perception, but its acceptance as ultimately real is an act of faith. In this theory we discard the concept of matter as inappropriate to modern science, but at the same time attempt to explain why it seems to exist in normal, everyday experience. There is a distinct structure of consciousness, we hope to show, that makes perceptual objects appear material on the macroscopic level, but that breaks down on the microscopic, or quantum level.

The approach in this theory is with consciousness as a first principle. We will consider all things, material and otherwise, to be manifestations of conscious experience, and we will presume the existence of nothing that is not conscious experience. There are no objects "out there" waiting to be perceived. All consciousness is composed of the same primal substance—that which appears "material" is dimensionally structured consciousness, and "nonmaterial" (thought and imagination) nondimensionally structured. Dimensions, therefore, are understood to be structures internal to consciousness itself, rather than external structures of a universe we live "in." This, of course, has tremendous consequences for the understanding of all physical phenomena. For instance, light is visual consciousness itself—it is not "in" space at all. The structure of space, in fact, is derived from light.

As a first principle, consciousness is impossible to define in terms of anything else. I will not attempt to say what it is as a whole, only to suggest that there is a unity of consciousness, and thereby of existence, that transcends differences between thought and perception, and also between one observer and another. But there is a certain structure to consciousness, the elements of which are definable. What I call *perceptual* consciousness is what you or I see now, or remember of what we saw yesterday. It is direct and immediate, and exists nowhere other than in direct and immediate experience. It does not exist "in" other observers. *Observational* consciousness, on the other hand, is indirect perception. It is what you or I *say* that we see. It is structurally related by means of dimensions to perceptual consciousness in that it is *potential* perception: I could always go where you are and see directly what you say you see. Perceptual consciousness is further divisible into five separate sensory realms, seeing, hearing, smelling, tasting, and touching, each dimensionally related to the others. Perception and observation are dimensional forms of consciousness, whereas thought and imagination are nondimensional.

This approach is radical, but by no means new. "Immaterialism," or "idealism," as this type of thinking is sometimes called, has been put forward many times before in varying forms and with varying degrees of success. Early attempts did not enjoy the benefits of modern physics and could not provide adequate alternative explanations for the apparent existence of material substance. Bishop Berkeley, for one, knew nothing of relativity theory, neutrinos, or the Heisenberg uncertainty principle, and presented nothing to replace or improve upon the concept of matter. We admire his insight, but cannot depend on the mere logic of his argument. We wish he were alive today.

I should say at this point that there are a wide variety of established philosophical meanings to the terms "idealism" and "materialism" that I do not wish to conjure. I use them only in a naive sense, one that is perhaps more understandable to the less philosophically schooled: a "materialist" believes that the world is really "out there" whether anyone is looking at it or not, while the "immaterialist" or "idealist" does not. There is a great deal of room for comparing the ideas presented here with those of Berkeley, Kant, Plato, and others, but to explore metaphysical parallels at this point would be digressive and distracting.

All previous attempts at immaterialism have suffered from the far more serious problem of the solipsism. If everything is consciousness, whose consciousness is it? If light is visual consciousness, is it in me or in you, or in both of us? Why do we see the same things? As far as I am aware, no immaterialist theory, before or since modern physics, has adequately addressed this problem. In this theory, I attempt to explain the apparent existence of separate consciousness "in" you and me and other observers in terms of the same dimensional structure of consciousness used to explain the apparent existence of matter. The strength of this theory is, I believe, its consistency over broad areas of application.

The dimensional structure of consciousness is fundamental to our everyday experience, yet has remained entirely unnoticed until the twentieth century. This is because human experience has been restricted to a range within which the concept of matter works quite well. We did not begin to explore the world of the very fast, the very small, the very massive, and the very distant until after 1900. There has been no need until now to notice the more subtle relation between consciousness and dimensions. Even now, this relation need not be noticed if we are willing to "shoehorn" what Planck, Einstein, Heisenberg, and others have said into traditional materialist categories.

In chapter II, the fundamental structure and content of consciousness are discussed, and the concepts of "image" and "realm" defined. Chapter III presents the concept of potential perception, which, I believe, is what we have always understood to be material substance. A potential is a "channel" of information, or a set of possibilities, only some of which are realized as actual sensory images. An infinite potential for sensory information is a dimension. In chapter IV, I discuss the workings of the dimensional structure of consciousness in macroscopic, or everyday experience. Mass is presented as an additional dimension, and each dimension shown to "correspond" to a realm of sensory perception. The intercoordination of dimensional potentials into "space-time" produces potential perceptions of a physical object in every realm: wherever and whenever an object is seen, it is potentially touched or heard, etc. This gives the object an appearance of independent, or "material" existence. "Dimensional interchange," or the rotation of axes in space, time, and mass, is shown to be what we experience as motion through space.

In chapter V, I discuss some of the enigmas that have puzzled physicists and philosophers since the introduction of relativity theory and quantum mechanics. It will be pointed out that each of the unusual physical phenomena presented here takes place at "dimensional extremes," (where objects are extremely small, rapid, distant, or massive), and that each involves the "act of observation," or the means by which physics becomes experience. This discussion is not intended to be comprehensive. Many readers will be familiar with these enigmas already, and it is my hope that those readers unfamiliar with them will study them elsewhere in order to more fully appreciate the need they create for a new metaphysic.

Chapter VI discusses the special significance of light. Light is visual information, consisting of trillions of minute tactile sensations, or photons. Light is therefore experienced on the quantum level in both the visual and tactile realms. Dimensional potentials lose their distinction: space, time, and mass merge into one another. Light is also shown to be an "intercellular medium," in that it serves in multi-cellular organisms as a channel of information between sensory and other types of cells. Retinal cells experience only the tactile value of photons, but give rise to multi-cellular "seeing" by encoding information in a space-time context. The "Quantum Screen" is suggested as a model of perceptual consciousness as a whole. Patterns of quanta on the screen, once they form a distinct "object identity," become the physical objects of macroscopic experience.

In chapter VII, I tackle the problem of the solipsism through the introduction of the "observational" realm of consciousness. Like the perceptual realms, this realm corresponds to a dimension that is interchangeable with other dimensions. Where we experience perceptual consciousness in the form of information from sensory cells, we experience observational consciousness in the form of information from observers. The existence of a dimension corresponding to observational consciousness makes it possible to propose a physical definition of life. The chapter is followed by some concluding remarks.

A major problem in expressing an idea such as this is the gap that exists between physicists and philosophers. It has been my experience that they do not listen well to one another, and often resent tampering across departmental lines. In speaking to the wide range of lay people who stand between them, I am unlikely to please either. But the gap is an unnatural one in the sense that it exists not in nature but in our attempts to understand nature. Quarks, neutrinos, and uncertainty principles have no regard for the departmental assignments under which they may fall.

The problem is one of language as much as of content. In an overly specialized age the means of communication within a particular field take on more importance than that which is communicated; the use of words and symbols becomes more a mark of fraternity than a conduit of truth. Style of communication becomes a matter of loyalty, and interdisciplinary work a matter of diplomacy. In this work I try to use as little specialized symbolism as possible. Mathematical symbols are avoided for the most part, and explained where unavoidable. A minimal exposure to high school physics and a willingness to put out some small effort should be enough to know what I am talking about. All of the examples from physics need not be understood thoroughly to understand the theory presented. In a similar way, I have avoided using terms and references favored by the philosophy profession. In both cases there will be losses in clarity, but it is to the more general reader that I wish to communicate.

Inevitably, and with reluctance, I have invented some new terms and concepts of my own. I have used descriptive terms with other common meanings, but I think this is better than concentrating on rigidly precise labels and watertight definitions. Terms too carefully defined in terms of other terms take on a life of their own and distract from the reality they are supposed to describe. "Perceptual" and "sensory" are used more or less synonymously to describe the direct experience of an individual with an object in space-time, even though they have slightly different meanings. "Observational," which has a meaning similar to the other two, is given an entirely new meaning: that of indirect, or communicated experience. The terms "potential" and "dimension" are used fairly explicitly. A "realm" of consciousness, however, is a "structured part of consciousness;" I do not think it can be defined more precisely without tedium. I use each term fully aware that I am merely barking at the thing itself, hoping the reader will notice it from within his own experience, and know its meaning. It is not my ideas I wish to get across in any case, but the realizations that have inspired them. If there is truth to this theory, it is not in what I have said, but in that at which I have aimed.

Additionally, it should be emphasized from the start that, though this theory denies the existence of an external material world, it does not view physical objects as illusory nor observers as merely mechanical or lifeless. To the contrary, the attempt here is to show that physics is a manifestation of life, and not life of physics.

II
IMAGES AND REALMS

Science cannot give us a complete picture of life because it can do nothing with the part of experience beyond its own realm of study. Most scientists admit that there is no way to understand "subjective" phenomena in any scientific manner.[1] What we actually experience within our own consciousness will never be experienced by others in any objective manner, remaining permanently beyond the bounds of science. Scientists try very hard, in fact, to keep the scope of their inquiry clear of subjective phenomena in order to avoid the taint of opinion or prejudice.

What do we do, then, with the very real yet untestable part of reality to which we cannot point? Must we admit that there is one reality for what "we" see, and another for what "I" see? Are we forced to conclude that there are separate and distinct realities that meet only at the surface between brain and mind? From a philosophical, theological, or psychological standpoint, this is entirely unsatisfactory. But it is just what we have been doing for the last three and a half centuries.

Within the last century it has become unsatisfactory from the scientific point of view as well. In physics, there is no longer a strict separation between subjective and objective. In each of the enigmas mentioned in the introduction and discussed later in this book, the "role of the observer" must be taken into account in order to understand the physics involved. In relativity theory for instance, an observer sees a rapidly moving object become shorter, gain in mass, and move through time more slowly only because his "frame of reference" is moving relative to the object. An observer moving *with* the object (in the same frame of reference) does not experience these dilations in space, time, and mass. In quantum mechanics, it is the act of observation itself that determines the outcome of an experiment. Extremely small particles pop into existence at indeterminate locations in space and time *only when they are observed*; where they are (or if they are) in between observations cannot be determined. In modern physics things do not just happen in an empty, dead universe—there has to be somebody, or some thing, observing an event for it to have physical meaning. This comes as a big surprise to physicists, who, until these effects were discovered, assumed that consciousness was an unnecessary appendage to the material world.

A clue to the relationship between consciousness and physics is that the enigmas mentioned above are not noticed in everyday life. They occur only at dimensional extremes: at extreme velocities, with extremely small particles, or in extremely strong gravitational fields. They happen where a space dimension is extremely large in relation to time (near light velocity), where space and time are extremely small (quantum mechanics), or, interestingly, where mass is extremely large (general relativity) or extremely small (quantum mechanics). Also, each of these effects involves distortions, discontinuities, or interconnections of space, time, or mass. (It is impossible, for instance, to know at the same time a subatomic particle's location in space and its momentum, or mass x space / time.) There is something fundamental, therefore, about the relationship between consciousness and the dimensions, something that we miss in the middle latitudes of space and time.

I will make a suggestion now as to what it is that we have missed. We assume that consciousness is *inside* of space and time. We think of it as a complexity of neural processes somewhere in our heads. If we turn this around and think instead of dimensions within consciousness, a continuity develops between what we call subjective and objective phenomena. Dimensions of space and time contain what we call "objective" phenomena: objective experience is dimensional, subjective experience is not.

There are problems with this, of course, not the least of which is that it does not make sense after a lifetime of assuming its opposite. But I will try to prove in this book that this is a better and simpler way to understand what we experience in modern physics, and in everyday life.

But what is experience, and what are dimensions? In this chapter I will try to show that experience consists entirely of what I call "images," and that dimensions are "potentials" corresponding to "realms" of consciousness.

Images

An image is a thought, a thing, a concept, a feeling, or an object; in fact, it is all things, physical and non-physical. It is the sound of a raindrop falling on the roof, or of an airplane in the distance. It is the picture of a place never seen, the memory of a taste experienced long ago, the touch of warm fur, and the pain of standing too long in one position. It is anger and lust. It is the Andromeda Galaxy, a moon of Jupiter, or a photon. It is a ball rolling down a hill. It is a cloud moving slowly across the sky that puffs up into large white billows, until raked by the wind, and combed smooth into thin wispy strands of smoke. It is an idea that ripples the mind.

Images are slippery and hard to catch. They are like fish in a river: if you reach down and catch one, it squirms and wiggles while you hold it, and slips back into the water.

Every thing is an image, and some things that are not things are images. Images are the sole content of consciousness, and constitute, for our purposes, ultimate reality. Reality consists entirely of images in their various forms.

Most scientists feel that images, while real in some sense, are no more than unreliable approximations of the physical world. For them my definition of image may be difficult to accept. I defend it only by saying that what they call the physical world consists of dimensionally-structured images. Some philosophers, on the other hand, particularly those of eastern religious traditions, may feel that ultimate reality lies beyond what I call images. They are right in a sense that I am not dealing with here. While I believe that the further progress of science depends upon a transcendence of the material world, I do not, in this presentation, attempt a transcendence of the normal consciousness through which we experience the material world. Eastern thought and practice present a fresh approach to the interworkings of thought, sensory information, and imagination, and an entirely new look at the western scientific tradition. I have found it extremely useful in understanding modern physics. The use of this approach is all I attempt here.

I should admit at this point that defining "image" as ultimate reality does away with the problem of saying what it really is. If it is ultimately real, and there is nothing else, what more can be said? This is exactly my purpose. I do not want to know what an image really is, nor how ultimate it may be, at least right now. What I am interested in here is the *structural relation among images* that creates the sense of physical reality we experience in everyday life. I want to know why, when I experience a visual image and a tactile image at the same time and in the same place, I think of something "out there," and why this sense of reality is distorted at dimensional extremes.

This theory, then, is an attempt to explain everything in terms of images without saying what an image is. The word itself is one I have had to select and weigh down with meanings, only some of which it can carry on its own. For rhetorical purposes, I have had to stretch and shape it, hopefully not too far beyond recognition. Also, as words are themselves images, none, including "image," is other than that which I wish to describe. My definition, therefore, is a tautology: I must use an image to impart an image of what I mean by "image." In any case, the word is the best available for my purposes in that it implies that all things, physical and mental, are essentially fleeting and ephemeral, and that imagination, while fundamentally identical with material substance, is in some sense more fundamental.

Objects are composed of images. You can touch an object because you experience a tactile image where and when you experience a visual image. The "object" is an intersection of images in space and time. It is this particular structure of consciousness, then, that creates the apparent existence of matter within objects.

There are images that are "real" and those that are purely "imaginary." The difference is that those we call "real" are experienced within a dimensional structure. Their dimensional context means that they are *potentially* experienced through any of the senses and also by any other observer. Images experienced subjectively are non-dimensional. The difference between "real" and "imaginary" is, I will try to show, the structure of the universe. Mental concepts differ from physical objects only by the context in which they are experienced, a context that we know as space, time, and mass. That they are both images, and therefore composed of the same primordial substance, I shall have to show. Or, I should say, I shall suggest that modern science has already shown.

But why has it been left to modern physics to discover a fundamental structure of consciousness when physics is not even interested in the structure of consciousness? Like Columbus on his way to India for gold and spices, physics has tripped over something it was not looking for in its search for grand unified theories and ultimate "building blocks" of matter. It has expanded the scope of human experience beyond its own conception of what is real. Before it began its voyages into relativity theory and quantum mechanics in the present century, the human mind was confined entirely to macroscopic dimensions. It is only with explorations inside the atom and beyond the galaxy that we have begun to peer around the edges of the dimensional world.

Realms of Consciousness

An image is never complete unto itself, but always relates to other images in some way. A particular image is similar to another in that there is a greater image that contains them both. A red house, for instance, bears some similarity to a green one in that there is such a thing as "house" that contains them both. The larger image serves as a means to locate and identify smaller ones within it. A pain in the jaw can be identified as a "toothache" because it is similar to other such experiences. The United States is a "nation" in that there are other nations like it. We know what an "automobile" is when we see one because we have seen so many others before. Conversely, every image consists of smaller, more fundamental images. A toothache is a combination of many separate "painful sensations," the United States is fifty "states," and an automobile is a particular arrangement of "bolts," "carburetors," and "seat covers." Conscious activity is a constant arrangement and rearrangement of images into other images, hopefully better, simpler, or more efficient ones. It is always a fluid process, and there is no perfect image or system of images containing all of experience. What I offer here, for instance, can be no more than a less imperfect arrangement than that I wish to replace.

One image that contains many others is what I call a "realm." A realm is a structured part of consciousness, containing a specific form of information. The "perceptual" realms are taste, touch, smell, hearing, and vision. There is also an "observational" realm of consciousness, consisting of information we experience through observers. It is similar in structure to the perceptual realms, and dimensionally related to them in the same way as they are to each other. There are six realms in all. The rest of consciousness is less clearly defined and structured.

But why are there six realms and not two or eight or twenty? Is there a metaphysical significance to this number? The number six seems to be accidental, like the number of planets in the solar system, or of chromosomes in the cells of a particular species. The number of realms is, however, related to our experience as humans (as opposed to plants or animals) at a particular stage of evolution. Plants experience two realms and animals anywhere from two to six. This has to do with the development of specialized sensory organs among higher animals, and of symbolic language among humans and some animals.

Realms are interrelated on the macroscopic level by coordinated dimensions. Three of these dimensions are spatial, and a fourth temporal (as demonstrated by Einstein in special relativity). I will make a case for mass, removed from the concept of matter, as an additional dimension. These first five dimensions are macroscopically distinct and measurable. The sixth dimension, that indicated by non-uniform acceleration, is less easily defined, but experienced in everyday life nonetheless.

The perceptual realms are also interrelated on the quantum level. The visual and auditory realms, for instance, consist of information that is *reducible* to tactile sensation, and thus to the tactile realm. Light is visual consciousness on the macroscopic level but at the same time *tactile* consciousness on the quantum level: each photon "touches" the retina as it becomes part of visual consciousness. The visual realm is, therefore, an outgrowth of the tactile realm. Extremely small images (objects approaching the energy of photons) are not exclusively visual or tactile (wave or particle) because the dimensional context within which they are experienced begins to disintegrate at this level. This is why we experience enigmas at dimensional extremes.

It is interesting that we notice the dimensional structure of consciousness only where it begins to unravel. We do not notice it in everyday life because it is everywhere.

Images and realms will pick up other names and associations as they wind their way through this discussion. In the next chapter we will discuss images in a single realm of perception; in this instance they become patterns of information in a dimensional potential. Later, as we discuss multi-sensory perception, images become physical objects in a multi-dimensional world. Finally, images become messages in "intercellular media" as we show that space dimensions arise only with multi-cellular consciousness. Images remain images in each case, no matter what their context, but it is their context that lends certain images the appearance of material substance.

III

POTENTIALS

Patterns, Probabilities, and Potentials

The purpose in throwing dice is not to know how they will come down. It is the uncertainty of each roll that gives the outcome its significance; the value shown on the dice when they land is without meaning except in terms of what it could have been. Before you throw two dice, all you know is that the number will be within the range "two through twelve." This is the "potential." If you roll an eight, you know that it could have been one of ten other numbers; the eight takes place within the potential "two through twelve." All bets, moves, and other contingencies that surround the game of dice are rooted in the foreknowledge that any particular throw can only be some whole number within this context.

But you may also know before you throw the dice that you are more likely to roll a seven than any other number. You may have learned this by watching repeated rolls of the dice at separate points in time. Each individual roll is totally random—it could be a six, an eleven, or a four—and is in no way affected by other rolls before or after. But some numbers are more likely than others. The dice have no way to "know" what they are supposed to roll, yet they seem to be guided by an invisible hand that causes more sevens and sixes and eights to show up than twos or threes or twelves. As some numbers are rolled more often than others, you soon learn that each individual roll has *probabilities* of being each number between two and twelve. Also, if you keep a record of each throw and make a chart showing how many times each number shows up, a distinct *pattern* shows up after many repeated rolls. The probability and the pattern are different ways of looking at the same thing. Each has meaning only within the context of the potential.

The probability for each roll looks like this:

numbers shown on dice:
    2   3   4   5   6   7   8   9   10   11   12

probabilities expressed as fractions:

$\frac{1}{36}$ $\frac{1}{18}$ $\frac{1}{12}$ $\frac{1}{9}$ $\frac{5}{36}$ $\frac{1}{6}$ $\frac{5}{36}$ $\frac{1}{9}$ $\frac{1}{12}$ $\frac{1}{18}$ $\frac{1}{36}$

Figure 1

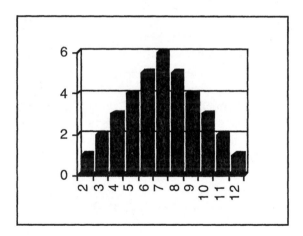

The pattern on your chart will look like this:
Figure 2

The probability and the pattern are immaterial and completely invisible at any single point in space and time. They are not in the dice, in the hands that throw the dice, in the table on which they land, or in the air through which they fall . But they do exist, and though they cannot be located precisely "in" the universe we know, no one would dispute the fact that they are very real physical phenomena. But what exactly are they, and how are we made aware of them?

Neither the probability nor the pattern is manifest in any single "event," or throw of the dice, but both are revealed after many throws. They are, therefore, a "wholeness" in the throwing of dice that is more than the sum of its parts, or the sort of "forestness" associated with every tree. They begin to break down when we look too closely at individual throws, and they disappear altogether between throws. We are made aware of them only when we stand back and view "throwing dice" as a whole, from a distance. They are, in fact, the sole reality of "throwing dice" in this sense.

But if they are both the "sole reality," how do the probability and the pattern differ? They are not exactly the same thing, despite the fact that they point to the same reality. In fact we do not really need both. If we are good mathematicians, the pattern is entirely unnecessary; the probability tells us everything we have to know. By looking carefully at the numbers we can comprehend the full reality of throwing dice without making a chart. The pattern on the chart is merely a visual aid to illustrate the same reality for those of us who are less mathematically astute. We create it by turning the potential into a *dimension*; we use equal measures of space to represent each number in the "two through twelve" potential. The pattern then represents actualities within the potential. With the help of the pattern, we can "see" the image "throwing dice" at a glance, and perhaps more easily and quickly relate it to other realities.

In plotting the pattern on the chart, a new meaning is created not by reinterpreting actual information from the dice, but by reinterpreting the potential within which the information exists. Our concept of potential experience shapes the meaning of our actual experience.

Let us suppose now that there is a creature of an entirely separate reality from ours who cannot see or hear, but is able, somehow, to perceive numbers directly. He cannot see us or see the dice, but as we throw them time after time, he is able to tell the numbers shown on them. After a while he figures out that they display a distinct pattern; he can "see" the more-or-less symmetrical curve shown in *Figure 2*. But the pattern, we have already said, is only manifest when the potential (two through twelve) is known; how does the creature know the potential? (He does not know how many dice there are nor how many sides each has.)

He learns about the potential through experience. He notices after watching repeated throws that the numbers showing up are never less than two nor more than twelve. This range of numbers eventually forms the background in his mind against which he interprets the significance of each roll. At the same time that he is establishing the potential, he begins to notice that there are more sixes and sevens than twos and elevens, and he is on his way to knowing the probabilities. By keeping track of many rolls, he is able to calculate the probability of a given number for each individual roll.

But he soon develops a quicker and more useful form of information. He begins to perceive many rolls of the dice as a single whole, rather than as separate, discrete events. The potential and the probabilities become unconscious: he starts to "see" the pattern, and not each roll. He develops, in other words, a two-through-twelve dimension in his mind as a context within which to interpret actual information.

The pattern is rough at first and difficult to "see." For a time there are more nines than eights, and two twelves and no elevens at all. But as he watches, the pattern becomes more symmetrical. The more he watches, the smoother it gets. In fact, the pattern becomes so smooth that he does not notice individual rolls at all, and he forgets that he ever developed a concept of dimensional potential. All he perceives is the pattern.

But the pattern becomes "orderly" only in the context of the potential; if there were no potential (if the number on the dice could be anything) there would be no order and our creature would have no idea what he was looking at. He would perceive no pattern at all. The potential is something he "realizes" through repeated experiences; he never perceives it directly. The potential remains invisible to him, and is therefore entirely *conceptual*; all that he actually sees is the number from each roll of the dice. But the potential also serves to *limit* the creature's experience; as he never perceives rolls of the dice less than two nor more than twelve, the potential becomes his "world."

He experiences the pattern even though it is not based on any kind of material substance. There is nothing "out there" that the pattern is "in." But neither is the pattern merely a product of the creature's mind; it is an actual perception "in the world," in the sense that the "world" is the potential.

The fundamental structure of his world is, therefore, entirely conceptual, though learned from actual experience. That his world is a dimension rather than the numbers "two through twelve" is due to the fact that he has learned to perceive the image "throwing dice" as a pattern rather than in terms of probabilities. If he were a better mathematician, he would not need the dimension as a visual aid, and his "world" would have remained purely numerical.

Sensory Information

In the world we know, what we see and hear are patterns of light and sound in space. Space is implicit in the information—it is the potential that lends special meaning to perceptual images, setting them apart from other types of experience. By its dependence on a background of unrealized potential, sensory information creates the universe of space and time.

There is a wide variety of definitions for "information" from the field of information theory, most of which utilize the concept of "entropy," or choice. The best "channel" of information is that with the most entropy, or the most possibilities from which to choose. A good channel of information would be a computer disk. There are so many possible arrangements of magnetic particles on a computer disk that an enormous amount can be said, or shown, in a very limited space. A telegraph key, on the other hand, is a fairly limited channel because each "bit" of information can only be a "dot" or a "dash," and it takes a long time to say relatively little. All of the information on the disk can be conveyed over the telegraph, but its limited randomness limits the efficiency of transmission. But that anything at all can be said through the telegraph or the disk is due to the fact that they present choices. There could be any other arrangement of magnetic particles or of dots and dashes. There has to be a background of random emptiness behind anything that is actually said. All information, in order to be information, depends on potential noninformation.

Dimensions serve in this manner as potentials for sensory information. They are near perfect channels in that they present nearly infinite randomness. Images in dimensions are patterns of sensory data, or finite ranges of points within infinite potentials.

According to Jagjit Signh, ". . . the message actually transmitted is a selection from a set of possible messages formed by sequences of symbols of its own repertoire. The communications system is designed to transmit possible selections, not merely the one that happened to be actually chosen at the moment of transmission."[2] The unchosen selections, or the numbers not shown on the dice, are, then, what turn the "actual" into information. If a dimension is a sensory potential, as I have suggested, it is the empty space surrounding a pattern that makes it part of perceptual, as opposed to conceptual, consciousness.

In a dimension, whether spatial or temporal, we have an infinite range of possibilities. This is the "entropy" needed as a background for the "order" we perceive in the form of sensory impressions. Actual sensory experience is "information" located at some point, or range of points, within a dimensional potential. Like the creature who never directly experiences the two-through-twelve potential as he watches numbers showing up on the dice, we never actually perceive dimensions.

But it is the dimensional context that compels us to think of perceptual experience as somehow more real than imagination. We feel that we are seeing and hearing something "in the world," something more than mere patterns of photons and air molecules. Space-time does this by making an actual pattern of information in any realm of consciousness potential in all others, thus giving each pattern an apparent life beyond the realm of consciousness in which it is experienced. It is this potential perception that is commonly understood to be material substance.

Einstein, though he remained a firm believer in an independent, material world, grappled with the problem of how we construe an "object" from sensory information:

> . . . we attribute to this concept of the bodily object a
> significance, which is to a high degree independent
> of the sense impression which originally gives rise to
> it. This is what we mean when we attribute to the
> bodily object 'a real existence.' The justification of
> such a setting rests exclusively on the fact that, by
> means of such concepts and mental relations between
> them, we are able to orient ourselves in the labyrinth
> of sense impressions. These notions and relations,
> although free statements of our thoughts, appear to

us as stronger and more unalterable than the individual sense experience itself, the character of which as
anything other than the result of an illusion or hallu-
cination is never completely guaranteed. On the other
hand, these concepts and relations, and indeed the
setting of real objects and, generally speaking, the
existence of 'the real world,' have justification only
in so far as they are connected with sense impressions between which they form a mental connection.[3]

Einstein does not, of course, identify his "concepts and mental
relations" as sensory potentials, much less as dimensions, but he does
admit that sensory information becomes "reality" by virtue of its conceptual context. A pattern of light, which could be "an illusion or hal-
lucination," becomes an "object" due to the "notions and relations" we
utilize "to orient ourselves in the labyrinth of sense impressions." The
"labyrinth" I take to be an accumulation of sensory information from
any or all of the five sensory realms, past and present, actual and
potential. Our orientation in the labyrinth I take to be our relation to
potential sensory information, or our position in space-time in relation
to the object. Where and when we see it determines where and when
we potentially hear, touch, smell, or taste it. The "reality" we attribute
to a sensory impression at any given location in space-time, I would
suggest, is due to its potential perception in every other sensory realm
at the same location. We confirm the "real existence" of an object we
see or hear by touching it at the same place. If we do not touch it there,
it is an illusion.

We have a great advantage, then, over the creature we have just
created. Where we have five senses by which to confirm the "real existence" of perceptual patterns, he has but one. His one dimension is his
world, and all he can perceive within it is a symmetrical pattern of
numbers shown on a pair of dice after repeated throws. A simple
world, at best. As he has no way to perceive this same pattern in any
other sensory realm, I doubt that he would attribute any sort of independent or "material" reality to it. Let us, then, give him more experience and see what happens.

Let us allow him to perceive numbers in other parts of the universe that we see as card games, lotteries, roulette wheels, and other,

more complicated, dice games. He becomes familiar with each game by watching it long enough to establish the potential and learn the probabilities. As he looks around him, he "sees" a wide variety of probabilities: some simple and symmetrical like our dice game, others more complex and convoluted. For card games, each card will be significant as a "location" within a potential of 52; probabilities will be vastly more complicated and depend on how and in what order "actualities" are revealed, or in other words, what the game is. Unlike the simple dice game, many of the new games (state lotteries in particular), have large "empty" stretches of potential that are never actualized. As he becomes increasingly familiar with the probabilities, he finds it useful, perhaps, to "dimensionalize" the potential as a visual aid. Each time a card is turned over, or the roulette wheel spun, the actual then fits into a dimensional pattern that he "envisions" in his mind. All of the patterns become smoother over time, as we have given the creature plenty of time to experience large numbers of the events that constitute each game. He has learned not to look too closely at individual events, but to wait until the patterns emerge. "Roughness" in a pattern is a problem only when he looks at too small a number of individual throws, spins, or turning over of cards.

With all of this new experience, have we broadened our creature's world? Is he able to develop a multidimensional world, such as ours, as a context within which to experience all this new information? Definitely not. Instead of broadening his old one-dimensional world into something larger, we have succeeded only in giving him several separate worlds. Each potential we have given him, even as it becomes a dimension, is essentially different; the two-through-twelve of the dice game is completely distinct from, and in no way coordinated with, the "ace-through-king" of the card games. The pattern he perceives in one game is in no way connected to that of other games. There is, therefore, no "confirmation" of the reality of a pattern in a separate realm, such as we experience in our world, and no inclination on the part of our creature to attribute existence to each pattern independent of perception.

The only way the creature could put all of his experience together into a single world would be to integrate each potential into a single context. He would have to learn to experience each of the games within a potential of the same fundamental structure; somehow he would have to experience a jack of hearts within the same sort of potential as rolling a six. In short, to live in a single, multidimensional world, he would have to learn to see each game in terms of the others, and to play more than one game at a time. This, I believe, is what we have learned to do with sensory information.

In our world, each type of sensory information is essentially different; light consists of photons, sound of air molecules, and taste and smell of chemical stimuli. But each "bit" of information is perceived within the same space-time context, so that each pattern of bits, no matter in what sensory realm it may be, constitutes an "object" that is potentially perceived in every other realm as well. We experience each type of sensory information within its own potential, but we have learned also to conceive each potential as a structurally identical and interchangeable dimension. Thus we have developed a single, universal context for all sensory information. The result is an integrated "picture" of reality.

The Limits of Dimensional Potentials

Let us suppose that one day, long after he has forgotten about watching us throw dice, our creature decides to investigate his world more thoroughly in order to identify its most fundamental building blocks. The patterns he perceives around him have by this time become so smooth that he has forgotten that they are made up of discrete numerical values that we see as individual rolls of the dice. He finds us again and watches us throwing the dice. After three or four throws, he notices that the pattern is so rough as to be barely perceptible: a three, two nines, and a six. What does this mean? Each throw seems to be entirely random, unguided by any pattern at all. What happens to the pattern in between throws? Where does it go? Knowing the pattern gives him some idea as to what the next throw might be, but he has no way of knowing for sure. It remains entirely "indeterminate."

Our creature, confused and depressed because his "world" seems to disappear when he looks at it too closely, begins to think back to when he first started "seeing" the pattern. He remembers, suddenly, that the pattern was created only when he began plotting the numbers on a chart, where the two-through-twelve potential was represented as a dimension. The dimension was something he came up with as a visual aid to help him understand the image "throwing dice." It was not in the dice, nor the dice in it; he learned to use it as a conceptual tool with which to interpret the results shown on the dice. But the pattern only emerged after many throws. To understand each individual throw, he would have to develop some other conceptual context.

Thinking back even farther, the creature remembers that before he began using the chart, there was a point at which he became aware that all throws resulted in numbers between two and twelve. As he watches individual throws now, he realizes that he can still use the two-through-twelve potential to give meaning to each result, even though it no longer works as a dimension. He can throw away the chart and instead calculate the chances of each throw being a six or a ten or any other number, and express it in the form of a probability relation. But this only helps him understand what is going on before each throw; afterwards, it is useless. Once thrown, the dice show a three or a four or a nine; there is no longer any probability that they will show other than what they do. The probability function "collapses" upon the actual event. But even after the throw, it is the probability that gives the result its meaning.

Our creature's dimensional world, then, exists only on the macroscopic level. On the "quantum level," or the level of the individual events that make up macroscopic patterns, there are no dimensions. Dimensions cannot therefore be fundamental structures of a world he lives "in," but conceptual tools within consciousness that give meaning to actual sensory impressions.

Our own world is similar to the creature's in this respect, differing only in that it is multidimensional. At the point where patterns become rough, dimensions begin not only to fade, but to blend into one another. According to the Heisenberg uncertainty principle, it is impossible to locate precisely at the same time both the position and the momentum of a small particle. The particle can be understood here as a small, and therefore rough, dimensional pattern, composed of a small number of bits of information. The smaller the particle, the rougher the pattern, as space and time no longer serve as appropriate potentials within which to interpret the individual events (throws of the dice) that make up the pattern.

The only way we can make sense of the behavior of individual particles is to develop a pattern based on observations repeated through time, under the same conditions (such as will be shown in the case of the two-slit experiment). The interference pattern develops on the photographic plate when the two holes are open only after a large number of particles have been allowed through (singly or all at once), in the same way that the pattern in the dice game shows up only after many rolls. Trying to determine "where" an individual photon is at some time during its flight from the light source to the photographic plate is like trying to determine where the pattern of the dice game is in between rolls. Space-time simply does not work as a context within which to interpret what is going on at each individual event.

If this is true, the question we are concerned with next is, why do the space and time dimensions of our world begin to disintegrate at the particular level that they do? Why do dimensions no longer serve as sensory potentials at the magnitude of the Planck constant, rather than at some other level?

We can begin to answer this question by noticing how space-time disintegrates. First, it does not fall apart all at once, but gradually, as distances, time spans, and particles become smaller and smaller. Second, time and space blend into each other as this happens, and also into mass. The smaller the particle, the more difficult it is to say what part of it is motion or location, and what part mass. (Physicists cannot determine a subatomic particle's "momentum," [mass x space / time] if they know its location at the same time.) And third, the breakdown of distinct space, time, and mass values happens at exactly the point where the momentum of photons begins to interfere with observation, that is, where the tactile value of each photon interferes with its visual value. This is the point at which individual photons become so few as to make visual patterns become too rough to be discernible.

There is no doubt that light plays the critical role in understanding both relativity theory and quantum mechanics. The enigmas found in both branches of physics can be understood only when light is understood in a completely new way. Rather than trying to understand why it causes problems on the quantum level we would do better trying to understand the role light plays in the creation of space and time dimensions on the macroscopic level. We will attempt this in a later chapter.

Finally, the reader may wonder how good an analogy the creature's world is to our own. I have not, after all, said exactly how the creature perceives probability patterns on the dice, only that he does so "somehow." What sort of sense organs might he have to accomplish this? I will not attempt to answer this question, nor the question as to how we perceive probability patterns within light and sound. But we do. Every shape that we see or hear in space-time is a pattern composed of trillions of energy quanta, quanta that could as well be composing something else. As they are a manifestation of consciousness, there is no process by which they become consciousness, and there can be no physiological model to explain it. We can explain relations between dimensional patterns in terms of causality, but there is no causal relation between information and consciousness. They are one and the same. If our creature has an unexplainable power to turn probabilities into perceptual consciousness, then so do we.

IV

THE MACROSCOPIC WORLD

In the everyday world, dimensions seem infinite and rectilinear, and objects appear to consist of material substance. This is apparent to such a degree that we assume it to be absolutely true. But we will find that this is not true at dimensional extremes, and therefore not true absolutely. I will try to show in this chapter that what we experience in everyday life can also be explained as a structural interrelation of the realms of consciousness, and that the concept of matter is entirely unnecessary.

To suggest that everyday life is understandable in terms of images and realms does not automatically make them better than matter as a metaphysical model. It may only make them as good. But I will show in a later chapter that, unlike matter, the model I am presenting also constitutes a workable metaphysical model for quantum and relativistic phenomena. It is in its broader scope that I will lay claim to its superiority.

Dimensional Coordination

A visual pattern is a physical object to the extent that we can touch it. Whether we actually touch it or not is unimportant; in fact, we touch only a very few of the visual images that we consider real. It is, therefore, the potential for tactile perception that makes a visual image "physical." We can extend this further to say that an actual image in *any* perceptual realm is a potential image in *every other* perceptual realm *at the same location in space and time*. If I hear something in the next room, I may smell, touch, taste, or see it there also. The concept of material substance, therefore, is derived from potential perception in each sensory realm.

We do not, of course, see one dimension and taste another. Dimensions are not actual perceptions; in fact, they are purely conceptual in that they are never experienced in any perceptual realm. Each dimension interrelates with each other dimension entirely outside of any realm. We do not see space-time; what we see are images that are coordinated in space-time with tactile, auditory, or olfactory images.

Quantification of dimensional relations is possible only due to this coordination of dimensions. Measurement is not possible within any single realm. A sound may seem "far away," a visual image "large," and a tactile sensation "heavy," but *how* large, far away, or heavy depends on where it is potentially perceived in other realms. Realms are intercoordinated through the projection onto each of essentially identical dimensional potentials. The concepts of "meter," "second," and "gram" derive from the interrelation of potential perceptions in macroscopic experience.

But why are they dimensionally coordinated? What causes us to see and hear and taste images at the same location? Normally, we think of material objects existing in space-time *causing* visual, auditory, or tactile images of the same shape and size when we happen to be looking, listening, or touching. There must be something external to perceptual consciousness, we assume, that gives us all these sensory images at once. But I am saying here that there is no causal process between object and perception—that consciousness has no "basis" in physics. There is nothing "out there" causing patterns of photons to match up with auditory and tactile patterns; visual information is perceived as a pattern in a potential that is already coordinated with every other sensory potential. Patterns are intercoordinated within each of the five sensory realms because the potentials in which they are perceived are intercoordinated. A physical object is nothing more than dimensionally coordinated patterns of sensory information.

Each point within an object bears a dimensional relation to the observer; it is this relation that exists in all five sensory potentials. Dimensional relations among points *within* an object (the size and shape of the object itself) have nothing to do with the intercoordination of sensory potentials. In other words, the length of an object has nothing to do with what it may smell like or how loud it may be. The length or width of an object merely shows the range within which perceptions are possible in other realms. If you pick up an object in the dark, for instance, you cannot tell by the range of tactile sensations what color it will be or what it will smell or sound like, but you can tell how long it will be in the visual realm, and that you may hear sounds associated with it anywhere from one of its ends to the other.

Each potential is a separate infinity, orthogonally related to each other infinity. We can visualize this fairly easily in the case of space dimensions: any amount of information can exist as a disturbance at a series of points in a single, straight line stretching off to infinity. Even if the line consists of points that can only be off or on (dots or dashes, perhaps), we can encode within it the entire works of Shakespeare or the Encyclopedia Britannica and still have an infinite length of empty points left over. If we wish to cross-index this information for some reason (say, we wish to have translations of every word of our original information into 197 different languages), we can create an infinite linear potential at every point on our original line. The result will be a two-dimensional surface stretching off to infinity in all directions. At every point on our original line where there is information, there will be additional information stretching off into the plane for some distance; where there was no information on the line there will be none on the plane. The result will be a finite two-dimensional pattern on an infinite two-dimensional surface. Most of the surface will be empty, but where there is information on any line running through the pattern there will also be information on any line perpendicular to it. We may not be able to say *what* it is, only *where* it is.

Without great conceptual difficulty we can add a third space dimension to our information pattern (creating an infinite three-dimensional space as a context), or even a time dimension. If we add time as a fourth dimension, our pattern will move as it passes through successive 3-spaces, and its context will be infinite space existing through infinite time. *When* there is information in our pattern, there will be information *somewhere* in each space dimension.

The essential equivalence of space and time was discovered by Einstein in his interpretation of the speed of light in the special theory of relativity. In essence, what he said was that space and time are the same thing, and that one second of time is equivalent to "c" meters of space, "c" being the velocity of light.

Four-dimensional shapes wandering through space-time are patterns of information that we experience as actual perception in one or more realms. When we experience actual coordinated patterns in more than one realm, they appear to be composed of some sort of "matter;" we feel that we are touching and tasting an apple "out there" in some independently existing spatial context. In fact, however, all we truly experience are tactile and taste patterns at the same location in space and time: there is no such thing as experience of a physical object independent of perceptual patterns.

We have discussed four dimensions of space and time, but mentioned five perceptual realms of consciousness, each of them "corresponding" to a dimension. What and where is the additional dimension? Also, what we have said so far seems to make sense for visual, tactile, and to a lesser extent, auditory perception, but gets a bit fuzzy when we talk about olfactory (smell) and chemical (taste) perception. It is hard to say "where" we smell something, and we can only taste something within the body. Also, are we not perhaps elevating these more primitive senses above their true place as minor and insignificant windows on the physical world? Can we really talk about the speed of light and the sense of smell in the same theory?

Dimensional Correspondence

We are interested here not in the actual contents of any sensory realm, but in its structural relation to every other realm. It is not important what we smell or what we taste; what is important is how smell and taste images relate to visual, auditory, and tactile images of the "same" object. We are more interested in the dimensional context of olfactory and chemical information than in actual tastes and smells.

The fact that taste and smell occupy equal space in human consciousness with other perceptual realms indicates an evolutionary past in which the contents of these realms played a much greater role than is the case now. If this is so, the particular structure of perceptual consciousness that we experience as humans is the same as that of other higher animals, though the contents vary enormously. The structural difference between human and animal consciousness is the existence of a nonperceptual realm that will be discussed in a later chapter.[4] For now we will concentrate on how individual perceptual realms correspond to space and time dimensions, and how each realm, as it evolves, adds an additional dimension to the world. To do this, we will have to dig deep into the evolutionary past, back to a time before there were eyes and ears and noses.

We begin by imagining, as best we can, what it was like to be alive in the early Precambrian era, before multicellular plants and animals had begun to appear. We can look back from where we are now, as humans, and say the Earth was like this and the sky like that, but this is not what I am getting at. We must imagine what it was like to *be* living things as they were at the time. There were as yet no "higher" realms of hearing, vision, and smell, only the chemical and tactile realms. We will have to be very simple, primitive, single-cell animals or plants. We can only touch and taste, and we will have to be what I will call "chemo-tactile" observers.[5] What would the "world" be like?

We have available to us two qualitatively distinct types of information. But how do we interrelate them, and what do they tell us about the world? More importantly, in what sense do we experience individual chemical and tactile images as "objects" in the world?

We can have no sense of being in space, because we can have no experience or potential experience beyond the cell membrane. There is no distinction, therefore, between "body" and "world." Without potential perception beyond the cell membrane, there is no world beyond the cell membrane. From the outside, of course, cells are observable as objects in a spatial context, but again, this does not tell us anything about what it is like to *be* a cell, which is what we are after.

But what about objects of perception? Do we not, as cells, experience things bumping into us, some of which we taste? Not exactly. What we experience are separate tactile and chemical images, at the same *time*. These we begin to think of as "objects," even though there is nothing "objective" about either image. The concept of "object" occurs when we associate a tactile with a *potential* chemical image: if we touch something we *could* taste it, whether we do or not. The survival value of tactile data is its coordination through time with potential chemical data. We know what we are getting into before we take something through the cell membrane.

The existence of time as a potential for chemical perception means that there is a "world" that we experience in a potential sense, and that there is more to the world than what we actually experience.

It is not the image itself that creates the world, but the dimensional background of the image.

Throughout evolutionary history to the present, all cells, plant and animal, including those in the human body, remain chemo-tactile observers. Multicellular organisms see, hear, and smell, but their cells do not. The experience of each cell is what contacts the membrane from the outside (tactile) and what passes through it to the inside (chemical). The cell membrane, in fact, is what creates the tactile realm as a sort of screening system for potential chemical stimuli.

It is interesting to note at this point that the plant kingdom, as highly developed as it may be in complex multicellular forms, has advanced no further than the chemo-tactile structure of perception. It has developed no "higher" sense organs, and therefore no spatial dimensions in its "world." Unlike animals, plants do not interchange dimensions, or move in space. The role of plants as observers in the normal sense is, therefore, questionable, even though there is no dispute that they are alive and sensitive to chemical and tactile stimuli.

Like single cells, we as humans do not taste everything that we touch, but do touch everything we taste. This means that there is no spatial context for chemical perception; we cannot taste anything at a distance. Even as vision, hearing, and smell evolve and become part of perceptual consciousness, only objects that are in contact with the body at the same time can be tasted. We can see or smell objects out in space, but not taste them. With the sense of touch and taste alone, therefore, and in the absence of the other realms, there is no such thing as space.

An *orderly* temporal sequence of tactile perceptions, on the other hand, may create a concept of externality and thus of spatial dimension. If you tap a paramecium on the shoulder in some rhythmic pattern for long enough, he may begin to sense that there is something "out there" causing the taps. His experience is limited to what he actually feels, but he may begin to imagine something like a space dimension in order to make sense of the order he experiences in the time dimension. In higher, multicellular animals it is, in fact, a temporal order of minute tactile sensations that produces the auditory and visual realms of perception and their corresponding space dimensions.

The sense of smell is a "higher" realm in that objects can be experienced through olfactory perception that are not in contact with the body. Olfactory information, therefore, exists in a spatial context. "What" is smelled is only potentially touched or tasted; it exists "somewhere" beyond the chemo-tactile world, somewhere that is not in, or on, the body.

Olfactory consciousness exists only within multicellular organisms and highly developed cell communities. Each cell within the organism is capable only of chemo-tactile experience; it is the organism *as a whole* that experiences olfactory images. Certain cells within the organism combine and specialize into olfactory organs, touching and tasting airborne or waterborne molecules that other cells in the organism do not. They encode this experience into information consisting of minute chemo-tactile sensations that other cells can "read." Nonolfactory cells experience this information in the form of *potential* chemo-tactile experience, that is, what they *would experience* in contact with the "object." No cell ever "smells" anything; olfactory cells experience actual, and nonolfactory cells potential, chemo-tactile sensations. Cells thus become aware of food, danger, or other stimuli that they never actually experience.

Olfactory information is encoded in such a way as to distinguish it from ordinary chemo-tactile perception; the entropic background of this information, or the "potential" within which the order exists, is qualitatively distinct from that of either tactile or chemical information. A sort of intercellular "medium" evolves, therefore, by which olfactory images can be communicated to nonolfactory cells. This medium is a single dimension of space. If we imagine ourselves to be olfactory observers evolving from chemo-tactile observers, our potential experience increases by a factor of infinity; our world develops a whole new dimension in which there are any number of objects that we did not experience before.

The olfactory realm serves the chemo-tactile realms the way the tactile realm serves the chemical realm: as an early warning system. Olfactory images are only potentially chemo-tactile, allowing an organism to screen objects before contacting them. Again, there is no actual experience of "object," only a sensory image within a dimension that indicates where information in other realms is located. An olfactory image tells us not that there is something "out there," only that there is potential sensory data in other realms that we were not aware of from within those realms.

But where exactly is it? All of us know that we cannot tell just where something is when we smell it. We cannot tell how far away it is or even what direction it is in; all we know is that it is out there somewhere. Without hearing or seeing it, we have to sniff around for some time before getting even the vaguest sense of its location, and even when we get close to it, we cannot say where it is until we do see it. An olfactory image, then, exists in only one space dimension: "elsewhere."

To determine the direction of an olfactory object, we must experience more than one olfactory image; we must "sniff around" until the smell increases. In experiencing multiple olfactory images we are making use of the time dimension, "interchanging" it with space. By moving about and comparing time-separated olfactory images, we "construct" additional space dimensions. But to know that we are creating additional space dimensions is again looking at an olfactory organism from the outside. From the inside, as true olfactory observers, with no sense of hearing or vision, we would never experience additional space dimensions even as we sniffed out an olfactory image. We would know only time and one dimension of space, and the world we live in would be limited accordingly.

The olfactory realm is only the first to use a spatial dimension as an information potential; hearing and vision are of the same basic type, and the evolution of each is accompanied by an additional space dimension. In the case of hearing, the new dimension allows for the directional location of auditory objects. As auditory observers, we can hear objects that we are not touching and also determine where they are, at least in terms of their direction. This is due to the structure of auditory information.

Hearing is reducible to the touch sensation of many millions of air molecules on the tympanic membrane. Individual cells in the ear cannot "hear" anything; what they experience is purely tactile. But "we" hear sound. "We," of course, are the experience of our cells as a whole, the media by which information is conveyed among our cells. "Sound" is a particular ordering of that experience. But order requires a background of randomness to appear orderly; a whole new set of possibilities must be created for the order to have meaning. A whole new "empty space" must come into being within consciousness before a series of minute touches can be experienced as sound. The potential is created, therefore, as a background against which the perception of sound becomes possible. That background is a second space dimension.

Sound takes the form of one-dimensional, or "longitudinal" waves. (Longitudinal waves can exist in one, two or three dimensions, but information is carried only in the direction of propagation. Displacement, or the disturbance caused by a longitudinal wave, is in the same direction as its propagation.) The direction of the source of a longitudinal wave can be determined if the receptor (the tympanic membrane) is itself extended in space, or if, as is mostly the case in living organisms, there is more than one receptor. Extension in space means that a time differential can be perceived in the arrival of a wave front. A sound wave arriving in one ear before the other means that the auditory object is closer to that direction, the magnitude of the time differential being proportional to the directional angle.

But the fact that a second space dimension is constructed from a time interval means that we are really "sniffing out" the direction of an auditory object the way we did for an olfactory object. There is a dimensional interchange involved between time and space, but where we must actively move about to create it for smell, it is implicit within sound. A *single* sound image contains two or more slightly time-separated wave-fronts; we do not "hear" a time separation, only the direction of the object. Unlike an olfactory image, a single auditory image is sufficient to determine direction.

But a single auditory image is not sufficient to determine distance. Extensive laboratory tests with blindfolded subjects listening to different types of sounds have shown that no matter how loud or soft an auditory object may be, observers cannot tell how far away it is.[6] They cannot determine, for instance, whether a loud object is really loud or just nearby, or a soft object merely distant. But this holds true only if the observer does not move, and if he cannot identify the object. If he moves, he is "sniffing out" the distance of the object with a second, space and time-separated auditory image, constructing a third space dimension through triangulation. And if he can tell how far away an object is because he knows what it is, he is using his past experience with the object, which consists of time-separated auditory images. Either way, he constructs the additional dimension in time; he does not experience it within the auditory realm.

The only way that we can experience the third space dimension directly is through the visual realm. With a single visual image, we can see objects "elsewhere," determine their direction, and also their distance. Having two eyes helps reinforce a sense of distance through added perspective, as does moving the head slightly, but the relative distances of visual objects are implicit in a single visual image. Light comes in the form of transverse, or two-dimensional waves. The displacement caused by the wave (its information content) is orthogonal to the direction of propagation: each photon vibrates at right angles to its path "through space." Each photon is its own wave, therefore, and has its own frequency or color, where each air molecule in a sound wave is only a small part of the wave as a whole. The relative distances of a visual image are thus determined by interposition. A visual image consists of a continuous pattern of photons of varying colors; any other pattern interrupting its continuity is closer to the observer. Relative distances are often determined in practice by multiple time- or space-separated visual images, but *can* be determined with a single visual image. As is the case with other perceptual realms, a new space dimension is added to the world not by the content of visual information, but by the background against which it is perceived.

Like sound, light is reducible to tactile sensation: photons are "no more than" minute touch sensations experienced by individual cells in the retina. The information potential, or intercellular medium, within which "we" read photons as vision is, of course, the third space dimension. Like sound, the dimensional context of visual information indicates where information in other realms is potential: where we see something is where we potentially touch it. As an intercellular medium, then, light is like sound or smell in that it informs nonretinal cells of potential chemo-tactile stimuli.

But light is unlike other intercellular media in one major respect. We cannot detect a *physical* medium for light the way we can for sound or smell. As will be shown in Chapter V, this is because space-time is itself derived from light and applied to other perceptual potentials. We define space-time in terms of light and all other perceptual information in terms of space-time. Light thus becomes the universal medium through which we experience all perceptual images. This universality of light-derived space-time makes all dimensions essentially equivalent and interchangeable, and creates a single multidimensional world instead of many one-dimensional worlds.

Before the development of the "spatial" realms of perception (seeing, hearing, and smelling), multicellular organization was possible only to a limited extent, and sensory organs were limited to relaying touch and taste experiences. Intercellular media conveyed chemical and tactile information, but these media did not create within the consciousness of the organism a world beyond its own body. There could be no sense of "otherness," no sense of being a body among other bodies. The leap into space was, therefore, a leap into an apparently separate consciousness over and above any single cell or group of cells. Organisms developed a sense of their bodies in space with other objects, and thus of their bodies as separate from other objects. Intercellular media conveying information between cells became the consciousness of the organism itself, the "wholeness" of the organism existing over and above any of its parts. The spatial realms constituted the organism as whole and indivisible, but also as separate from the rest of the universe.

Mass as a Dimension

The olfactory, auditory, and visual realms each correspond to space dimensions, and the chemical realm to time.[7] The tactile realm must, then, correspond to some as yet unmentioned dimension. But how is this possible if the world we live in is four dimensional and complete in itself? There is simply no room for more space or time. Where, then, do we find another dimension?

If we were to find it anywhere, or anywhen, it would not be a dimension. An additional dimension could not be a mere appendage to space-time, but would have to be an infinite range of new possibilities. There would have to be an infinite number of possible space-times within *it*. As the dimensions we already know are not external structures of an independently existing world that we are "in," we do not have to think of an additional dimension or dimensions as something we are "in." Space and time, we have seen, are information potentials; what we should look for is another potential in which actual images are located, and their location should tell us something about potential tactile perception. If we can make measurements within it, the way we can in space and time, and we find that it is interchangeable with them, it may be the dimension we are looking for.

We have already associated the body with the tactile realm of perception. But do we mean actual or potential tactile perception? If there are no spatial realms, as in the case of a chemo-tactile observer, there can only be *actual* tactile sensation. In the absence of seeing, hearing, or smelling, there is no such thing as "potential" tactile sensation, and no body in space-time waiting to be touched. Tactile sensation remains, but it has no special context apart from the rest of the world; the "body," therefore, does not exist at all. But with the development of the spatial realms, the body comes into existence as *both* actual and potential tactile perception in separate realms: in the tactile realm, the body becomes actual perception, and in the visual, auditory, and olfactory realms, the body becomes potential perception. In the tactile realm, what we actually feel is the body and what we potentially feel is the inertial mass of objects around us; in the visual realm what we see of the body (a body part) is where we experience potential tactile sensations. Where we see a finger or a leg, for instance, there is potential tactile sensation. The body is, therefore, an actual tactile image that becomes potential in all other realms.

But if the body is an actual image in the tactile realm, what is the tactile potential? There must be a larger context of potential tactile perception within which the body finds itself as a pattern of actual information. But it is not a space dimension. Putting the body in space coordinates touch with smelling, hearing, and seeing, but does not provide a dimension corresponding to the tactile realm. The body cannot feel, or potentially feel, the space that it is in. Even when we put the body in space *and time* and allow it to interchange one space dimension for time (allow it to move with constant velocity through space, as we will see below), it feels nothing. The only time it feels anything is when we *accelerate* it, or change its velocity. Whenever we accelerate the body the entire tactile realm is activated: we feel a kinesthetic sensation throughout all parts of the body and the magnitude of this sensation is in proportion to the magnitude of acceleration. This motion is in the *second time dimension*, as we measure acceleration in terms of meters per second *per second*. The "g" force of acceleration is the tactile potential itself. It is coordinated with other potentials as a second time dimension.

We feel the "mass" of an object in terms of its weight, or its resistance to acceleration: the more massive the object, the less it accelerates when we push it. We establish a standard for measuring mass through the use of a known force such as a spring, but it is only in terms of acceleration, or a second time dimension, that mass is apparent at all. For this reason, we will call this second time dimension the "mass dimension" and consider all physical objects to have a location in it as well as a location in space and time. The location of an object in the mass dimension (its inertial mass) does not indicate what it may feel like any more than its location in space indicates what it will look or sound like, but it does indicate how it will disturb the body as a whole if contacted. Rather than trying to find "matter" in physical objects, we will instead find physical objects at a specific location in "mass." If an image has no mass value, it cannot be touched.

But how will a mass dimension work in traditional physics? Physical knowledge has always been based on measurements of space, time, and mass and expressed in terms of the fundamental units "meters," "seconds," and "grams." Complete physical knowledge of a macroscopic object (its "primary qualities"[8]) consists of its length, width, and thickness (all space dimensions), its velocity (space/time), and its mass. All physical concepts such as force, energy, pressure, momentum, even temperature, are made up of some combination of these units. As a dimension, mass will work in physics as it always has.

An object's location in mass is thus analogous to its location in space: it is always in relation to some other object or to an observer. But an object's *center of mass* is analogous to its intrinsic spatial dimensions, or to its shape in terms of length, depth, and width. Center of mass and other intrinsic dimensions determine an object's identity, not its location; identity, including center of mass, may remain despite changes in location. If we think of an object as a range of points in space-time-mass, the location of *each point* in relation to an observer determines its potential perception in all five sensory realms; the relation of points within the range determines object identity.

Meters and seconds were unified as "space-time" in the special theory of relativity; here we are adding mass to make "space-time-mass."

What is experienced and measured does not change, what changes is the understanding of what is experienced. Instead of "matter" in "space-time," we have images in "space-time-mass."

In the general theory of relativity, Einstein himself points toward mass as a dimension when he speaks of gravitation as a "curvature" of four-dimensional space-time in the vicinity of massive bodies. But into what is space-time curved? If we can speak by analogy of a curved, two-dimensional surface, we can better visualize the problem: A two-dimensional surface such as that of the Earth may be so large that we do not notice its curvature by any local measurements. The shortest distance between two points is always a straight line, and the angles of a triangle always add up to 180 degrees. But if we start making measurements over extremely long distances (over vast stretches of ocean, for instance), distortions become apparent: the shortest distance between two points on our surface becomes a curved "geodesic," and the angles of a triangle add up to more than 180 degrees. Without experiencing it directly, we can deduce the existence of a third dimension into which our surface is curved. In the same way, when we find that "intervals" between "events" (distances between points in four dimensions) are distorted near extremely massive bodies, we may deduce that our four-dimensional "surface" is curved into a fifth dimension.

Einstein was intrigued by a five-dimensional idea based on his own work and introduced by Kaluza and Klein.[9] Light was considered a vibration in an additional dimension orthogonal to space-time. As the energy of a photon is proportional to its value in this dimension (according to Einstein's formula $E = hv$), which is the photon's impact on the retina, the additional "per second" dimension discovered here must be the same as the "per second" dimension revealed on the macroscopic level as mass. Einstein and others showed great interest in this approach initially, but later cast it aside for lack of apparent physical manifestation. The curvature of four-dimensional space-time into a 5-space made good logical and mathematical sense, but was rejected because the real world was thought to be limited to four dimensions. "Matter" was still in "space-time." The same idea could be revived, I believe, and combined with the idea of the tactile reduction of light.

Even Descartes suggested two and a half centuries earlier that mass might be considered a dimension. "By dimension I understand not precisely the mode and aspect according to which a subject is considered to be measurable. Thus it is not merely the case that length, breadth, and depth are dimensions, but weight also is a dimension in terms of which the heaviness of objects is estimated."[10]

An advantage of the five-dimensional world view is that neither the Mach Principle (discussed in the next chapter) nor absolute space are necessary to explain inertial mass. Modern physics has no good explanation for how or why mass "resists" acceleration, much less why we feel acceleration throughout the body. What is it about space that "holds back" massive objects? And why can we feel acceleration but not feel constant velocity? It is simpler to say that at every point in space-time there is an infinitude of possible locations in the second time dimension. The acceleration of an object is its motion in five dimensions; the second time dimension appears "foreshortened" in the first time dimension much the same way a third space dimension often appears foreshortened on a two-dimensional surface. The location of the object in the mass dimension is revealed by its rate of acceleration when subject to a force of known value.

Dimensional Interchange

Five separate but intercoordinated sensory potentials are what we call the "real" world. Dimensional images are objects in this world: actual patterns in one realm and potential in all realms. Nondimensional images are "imaginary," having no place in the "real" world. But if only part of our experience is dimensional, why do we feel that we are "in" this world?

We have spoken of space as a relation to the body, and not as an absolute, external structure. The dimensions we have discussed so far are "elsewhere," "direction," and "distance"—each of them in relation to the body—not length, height, and width, as would be the case if we were viewing the body from a distance. This is because the perceptual realms alone (without the observational realm), do not allow us to experience the body from a distance. The body remains the center of the universe and therefore the origin of dimensional axes. A rotation of these axes, or "interchange" of one space for one time dimension, is experienced as a uniform shift in time of the spatial relation of all objects in the universe to the body, or, what is the same thing, a uniform motion of the body in space. The body moves through space in the opposite direction.

This does not mean that consciousness is in, or moves in, space-time, only that there is a shift in perceptual potentials in relation to the tactile potential. The location of images shifts in each realm. In the visual realm, the tactile potential moves into contact with visual or auditory images, and they become touchable. It is through dimensional interchanges of this sort that potential perceptions can become actual in a meaningful or orderly manner. It is the means by which we manipulate or attempt to control experience, or, in other words, what we call free will.

To be interchangeable, dimensions must be essentially identical. This has already been shown to be the case with space and time, but what about mass? In what manner is mass interchangeable with space, or with time? The mass dimension, we have said, is the second time dimension and corresponds with the tactile potential. The tactile potential is the body in the visual realm and the second time dimension for the tactile realm. We *see* the body (or a part of the body) as an object "in" space-time where there are potential tactile sensations, but we *feel* it as a pattern in space-time-mass, or as an acceleration. When we see all objects in the universe accelerating past us at a constant rate, we at the same time feel a universal tactile sensation of a magnitude proportional to the rate of acceleration. This is experienced as an acceleration of the body through space in the opposite direction, or as a rotation of axes in space, time, *and mass*. This is an interchange of the mass dimension with space-time; if it involves one space dimension, it is linear acceleration, if two, angular.

We experience dimensional interchanges regularly in everyday life. When we get up and go somewhere, or just turn our heads, we feel accelerations throughout various parts of the body as we see and hear objects shifting their dimensional relations to us. When riding in a car at a more or less constant velocity, we feel bumps in the road as minor upward accelerations, and curves as centripetal accelerations. In fact, we experience a constant rotation of axes into the mass dimension in the form of gravitation: we always feel the "pull" of gravity in our bodies, and see unsupported objects accelerate past us to the ground. The only way we could experience life without dimensional interchanges would be in interstellar space or free fall in a gravitational field.

Rarely, if ever, do we experience pure or complete dimensional interchanges. We can never interchange just space and time, as there are no roads without bumps and curves, and we can never experience the body as a whole in a perfectly uniform manner. Whenever we experience a part of the body, we also experience the body as a whole, and whenever we experience the body as a whole, we always experience a part of it. There is never a complete distinction between whole and part. When we touch something, for instance, no matter how small it may be, the whole body is accelerated in the opposite direction to some extent. Conversely, when the body is accelerated, it is always in contact with some object. We feel the back of our chair as a jet roars down the runway, or we feel the floor beneath our feet as we stand against the Earth's gravitation. We may resist acceleration of the body as a whole by pushing against some other object, as when we brace ourselves to catch a football, and in so doing we may experience no *overall* shift in the second time dimension. But we do experience two or more shifts that counterbalance.

In everyday life we usually counterbalance minor disturbances in the body with equal and opposite accelerations, shifting our weight in order to keep our balance. This way, when I hand you a pencil, you push slightly against the floor in my direction so as to avoid drifting aimlessly down the hall. We may experience great disturbances within the body, therefore, without an overall interchange with the mass dimension.

The body, then, acts as a sort of interface between the fourth and fifth dimensions. In four dimensions, it is a shape in visual space-time outlining the limits of potential tactile sensation. In five dimensions, it serves as a point of origin of dimensional axes. Whenever there is an overall disturbance in the body, we experience a rotation of axes in the mass dimension.

The concept of dimensional interchanges rounds out our picture of macroscopic experience without material substance and without consciousness in space or time. Everything that we experience in everyday life can be explained in terms of the metaphysical foundation outlined above. But this makes it equal to the material picture of physical reality, not better.[11] Our picture will prove better only if it shows greater scope, that is, if it can also explain some things that the material picture cannot. In the next chapter we will explore what happens at the limits of macroscopic experience, or at dimensional extremes, where the potentials that carry sensory information become distorted. It is where space, time, and mass become unraveled that they reveal the structure which holds them together.

V

THE ENIGMAS OF MODERN PHYSICS

Dilations in Space, Time, and Mass

At the turn of the twentieth century the physical world was well understood. Enormous advances in chemistry and astronomy had confirmed the fundamental laws of nature as stated by Galileo and Newton, and new work in electromagnetics was bringing together previously diverse fields. Material objects followed straight lines in space unless disturbed by external forces; actions had equal and opposite reactions; and the sun pulled the planets around with inverse square laws. Matter was composed of trillions of invisible atomic particles that could combine in any number of possible forms. Light traveled in waves[12] through space, bouncing off matter that lay in its path, and entered the eye, causing us to see. All this seemed to underline the fundamental order and unity of creation. The world was hard, cold, rational, and without need of life; life itself was no more than an especially complicated form of matter, an afterthought to creation. The future of science lay not in new fundamental discoveries, but in confirming what had been revealed already, by filling in blank spaces and resolving any remaining details. This was classical physics.

But toward the end of the century some details developed that could not be worked out within the metaphysical limits of classical physics. The most glaring of these was that nobody could find anything that "waved" when light passed through it. Water molecules moved when a water wave passed and air molecules moved when sound passed: what, then, moved when a light wave passed? The confidence that some medium of light transmission would be found was so great that it was given a name before it was discovered: the "luminiferous ether." Nobody knew what the ether was or how it carried light and other forms of electromagnetism, but it had to be there. You cannot have waves of nothing.

To detect a medium you have to move something through it and see how that motion is affected. If you can row a boat at five miles per hour upstream and at fifteen miles per hour downstream, you can tell that the boat is moving at ten miles per hour relative to the water, and that there is a five mile-per-hour current. You have detected the river as the medium carrying your boat. In a similar manner, you can measure wind velocity by measuring the speed of sound upwind and downwind. Sound waves travel faster downwind because the air molecules of which they consist are moving downwind. If we want to find a medium for light, all we have to do is move a light bulb relative to that medium, turn the light on, and measure the difference in light velocity "upstream" and "downstream."

But we will have to move something very fast in order for its velocity to be noticeable in relation to the 186,000 mile-per-second velocity of light. This might sound difficult, but it is really quite easy, as we find ourselves already riding through space on an object that goes quite fast. All we have to do is to measure the speed of light in the direction of the Earth's motion around the sun and compare this six months later to light velocity in the opposite direction. Albert Michelson performed this very experiment several times in the late 1800's and early 1900's with an extremely sensitive instrument that he had invented called an "interferometer." But to everyone's surprise, he could detect absolutely no difference in the speed of light in any direction, at any time. There seemed to be no medium for light waves.

This was serious. Most scientists thought that there had to be something wrong with Michelson's instrumentation, or with his laboratory, or that the Earth "dragged" the ether around in its orbit, or that there was some other as yet undiscovered explanation. In 1905 Albert Einstein resolved the difficulty by stating in the special theory of relativity that light moved at the same velocity relative to *everything*. There was no such thing as the ether. That is why no changes in the velocity of light can be measured, no matter how fast the measuring instrument may be moving. Even if the Earth were moving at two-thirds or nine-tenths the speed of light, light would always be moving at the same speed relative to it, and to everything else. There simply was no medium that light waves moved through.

This violated everything Newton, Galileo, or anyone else had ever said about relative velocities. If I am in a train moving at fifty miles per hour, and I roll a ball up the aisle at five miles per hour in the same direction, it moves at fifty-five miles per hour relative to the ground. Everybody knows that. So if light travels at exactly 186,282.0000 miles per second, and I turn on a flashlight and point it towards the front of the train, (still moving at 50 miles per hour, or .0138 miles per second) the light should travel at 186,282.0138 miles per second in relation to the ground. But it does not. It travels at 186,282.0000 miles per second relative to the train *and* the ground. This, of course, will not work in the world we think we know, and something has to give somewhere. Einstein, to keep the velocity of light constant for both the train and the ground, had to make *time* pass at different rates for each of them. As I look out the window at you on the ground measuring the speed of light from my flashlight, I notice that your watch is running *slower* than mine. You and all the buildings and trees and molecules around you do not experience the passage of time the way I do. You, in turn, look at my watch, and see that it is running *slower than yours.* Because we each measure the velocity of light in terms of meters per second, the difference in our seconds just makes up for the difference we should be measuring in velocity. The passage of time is relative to the reference frame from which it is observed.

I also notice that the meter stick you use to measure distances is shorter than mine when you point it in the direction of the train's motion. Each of your meters is shorter than mine! As you measure objects around you, you can fit your meter stick along them more times than I can mine, and so you think they are longer than they really are, at least as I measure them. But you think *my* meter stick is shorter than yours! Also, I measure your *mass*, and the mass of all objects around you to be greater than you do, and you measure my mass and the mass of the train to be greater than I do. If we stop the train and get together to talk it over, we find that things are back to normal; we each measure the same values in space, time, and mass. But each of us could swear that the other was wrong while the train was in motion.

Neither one of us is right in an absolute sense; I had my reality and you had yours. Everything in my "frame of reference" (the train) seemed normal to me and everything in yours (the ground) seemed normal to you: it is only when we looked at each other's frames of reference that distortions appeared. And those distortions were very minor for a train traveling at fifty miles per hour. At that speed, the difference we noticed would be a fraction of a second in a million years or so, a millimeter every other light-year, and the weight of a proton to a bull elephant: nothing to be overly concerned about. In fact, we do not notice these things at all in trains moving so slowly relative to the speed of light. They are very much there, but much too small to be noticed, even with the most sensitive instrumentation.

These "dilations" in space, time, and mass occur according to the following formula:

$$\gamma = \frac{1}{\sqrt{1 - \dfrac{u^2}{c^2}}}$$

where "γ" is the amount of dilation, "u" the velocity of the train, and "c" the velocity of light. If "u" is only 50 miles per hour, and you remember just a little high school algebra, u/c is near zero, "γ" is almost exactly 1, and the effects are so small as to be immeasurable.

If we want to notice them, we have to pour on the coal and move the train at several million miles per hour. If we bring "u" up to 99.99% of the velocity of light, u/c will be .9999, $\frac{u^2}{c^2}$ will be .9998, and "γ" will be 70.71. Now your meter stick will be less than a centimeter and a half long, your watch will take over an hour to show the passage of a minute, and you will weigh as much as that bull elephant we were talking about. This I will notice.

Now 99.99% of the velocity of light is a "dimensional extreme." Velocity is not itself a dimension, but a relation between space and time dimensions; it is measured in terms of meters per second. When we try to go too fast, we put too many meters in each second, and dimensional dilations result. Dimensional extremes are where we *notice* dilations, but are not the only places they exist. They exist in more normal circumstances, but reduced in magnitude. Even as I walk past you down the street, there are small differences in the dimensional relations each of us experiences. The differences may be negligible several million times over, but the fact that they exist at all means that there is something unstable about our idea of the world we live "in."

We assume, you and I, that we live in the same world. Differences between what you see and I see can be worked out by factoring out our separate perspectives in space and time. You see a different angle on things because of where (and when) you are looking from; if I were where you are, I would see what you see. One way of understanding special relativity is to say that the "γ" factor is just another matter of perspective; you see *me* traveling at 99.99% of "c" in the opposite direction and think *I* weigh as much as a bull elephant. If we factor out all the "γ's," we can live in the same world again. But it is not the sameness about your world and mine to which I object; later, I hope to show that in the observational realm we *can* factor out separate perspectives, including the "γ's," and have the same "world." What I object to is either one of us living *in* it. To get us out of the world, we have to get light out of space-time.

Let us look again at the "constant velocity of light." You think my meter stick is only a centimeter and a half long when we are traveling relative to each other at 99.99% of "c" because we both measure a ray of light to be traveling in space at the same velocity. But how can it be "traveling in space" at the same rate for both of us? Space cannot be an external world structure if it allows itself to be distorted for the convenience of each observer. How can we each carry around our own velocity of light? Are we each carrying our own sense of space? We define space in terms of light, and now we are trying to measure light in terms of space! Even the straight line we are traveling relative to each other is defined by the path of light "through space." Also, "c" is more than a velocity: no object, according to the relativity theory, can ever travel faster than "c:" *empty* space holds everything back to less than "c" meters per second. How? And when multiplied by itself, "c" also becomes, according to the special theory, the link between mass "m" and energy "E":

$$E = mc^2.$$

What does this have to do with light?

Light is a great deal more than a wave (or a particle) traveling through space. It is more even than visual consciousness. The structure of light, (the relation between space and time in visual consciousness) is the structure of every other perceptual realm as well. We "envision" objects in a space-time modeled after light even if we do not see them. We perceive objects that we hear, smell, taste, and touch where we *would* see them. "c" is the structural relation between space and time within light, but also the structural relation between space and time itself.

Einstein showed that space and time are equivalent and interwoven in terms of "c" into "space-time." One second of time is equivalent to "c" meters of space. In the same way that an extremely small time interval is "heard" as the direction of a sound wave, an even smaller time interval is "seen" as distance. One second old is the same thing as "c" meters distant: a star seen ten light years in the distance is where it was ten years before. Even a chair ten meters across the room is where (and what) it was 10/c seconds ago. This is a built-in limitation of light as an information medium:[13] meters and seconds cannot be completely separated because they are fundamentally the same thing. Objects moving close to "c" meters per second in relation to an observer expose this limitation.

But why is light limited? Why is "c" not infinite? Light is an imperfect medium because it consists of patterns of tiny tactile sensations, or photons. As long as these patterns are not too small, too massive, or too fast, the medium works well and the patterns look smooth. But at dimensional extremes, the "graininess" of the medium begins to show. Photons are like the tiny ink dots that make up a photograph in your morning newspaper. If you look at too great an angle or too closely, the picture itself becomes distorted or disappears altogether; all you can see are the dots. The structural relation between the dots (their "c") is their distance from one another; the smaller the distance, the smoother the image. The "c" for photons is so good that we do not notice in everyday life that they are really tiny dots on the retina.

The dimensional dilations of special relativity are distortions of the sensory potentials within which images are perceived. To understand them, we must understand light not as a phenomenon in space but as visual consciousness itself. Space is the context within which visual information becomes a realm of consciousness.

The Heisenberg Uncertainty Principle

At the turn of the twentieth century it was also thought that dimensions were smooth and continuous. Objects existed at precise locations in space and exact moments in time, and consisted of precise amounts of matter. Measurements of dimensional values were, of course, limited by available instrumentation, but as instruments became more accurate, dimensional values could be determined with ever increasing accuracy. There was nothing built-in to the world that would limit the number of decimal places you could add to a measurement. If you had the right tools, you could say exactly where a particle was no matter how small it might be, how much it weighed, or how it was moving. Most people still assume this to be the case, as dimensions appear perfectly smooth and continuous in everyday life. But in the year 1900 Max Planck discovered that energy is not perfectly smooth and continuous, but comes in extremely small but discrete bundles, or "quanta." He surprised himself as much as anyone else by so revolutionary a discovery, and without meaning to, invented a whole new branch of physics known as "quantum mechanics."

The Planck constant "h," relates the energy "E" of a photon to its frequency, "υ," by the following formula:

$$E = h\upsilon$$

Each photon of light is a quantum,[14] or indivisible unit of energy. But the amount of energy in each light quantum varies with its color, or frequency: the higher the frequency, the greater the energy. A purple photon is a higher-energy quantum than a red one. Quanta can and do exist in forms other than light, but it is only light quanta that can be detected individually.

The reason we do not notice quanta in macroscopic experience, and that they were not discovered until fairly recently, is that they are so small. The Planck constant is equal to 6.6 x 10^{-34} joule-sec. This gives an idea of the "size" of a quantum: the "66" would appear 34 decimal places to the right of zero. This would be 66 ten-thousandths of a quintillionth of a quintillionth; you would not know if you stepped on one.

What this means is that when objects move, they do not pass smoothly through space, but "jump" from one quantum to another.[15] When you throw a baseball, for instance, you think of it as a distinct object of measurable size and shape, sliding parabolically through empty space. That is what it looks like, sounds like, and feels like from a macroscopic point of view. But if you were able to slow the baseball down and look at it through an extremely high-powered microscope, you would see that it is actually moving in many trillions of discrete increments, or jumps, much the way Pac Man moves across a video screen.

If you look very closely at Pac Man, you will notice that he is really made up of tiny lights on the screen; as he moves, new lights light up ahead of him and old lights shut off behind him. It is just the pattern of lights that moves, not the lights themselves; we think of him as a "thing" moving across the screen, but he is really a discontinuous pattern of tiny points of light. From up close we can see that the pattern jumps from one set of lights to another, and that Pac Man is really moving in a very jerky manner. From farther back, however, he appears to move more-or-less smoothly. The farther back we are, the smoother he is. In the case of the baseball, we are so far back that it appears to move perfectly smoothly.

The baseball is a discontinuous pattern just like Pac Man. It jerks from one quantum to another, turning on trillions of photons ahead of it and shutting them off behind. (If we want to catch the ball, the location of the pattern tells us just which dimensions to interchange to bring it into the tactile realm.) Like Pac Man, the baseball is information; it is an orderly pattern against a background of potential randomness. We see a "thing" arcing through "empty" space only because the pattern stays the same when it could disintegrate into total disorder. We did not know this about the baseball until Planck discovered that there was a "quantum screen" against which it appears.[16]

Pac Man's screen is much more limited than Planck's. It has only three dimensions: two of space and one of time. (Pac Man has neither depth nor mass: he is flat and can accelerate instantaneously.) Planck's quantum screen has five dimensions. (Energy is expressed in terms of grams x meters x meters / seconds x seconds.) The Pac Man screen is also much coarser than Planck's; the tiny lights on a video monitor are much bigger than quanta, and enigmas show up at macroscopic magnitudes.

But this makes the analogy work: Pac Man's screen is coarse macroscopically the way Planck's screen is coarse microscopically.

To look at the quantum screen itself, we need to use objects much smaller than baseballs: tiny particles like protons, electrons, and pi-mesons. This will be like making Pac Man smaller on his screen. What happens? You take him down to half size and you can still make him out, only less clearly. You take him down another notch or two and he gets fuzzy; details disappear and lines appear less distinctly. If you get him down as small as three or four lights on the video screen, what is left? He can still move around, but what is distinctly "Pac Man" about him, and where is he when he is between one light and another? To what extent does he exist *in between* lights on the screen, and "where" does he exist "when" he is on his way between one light and the next?

Nearly a generation after Planck, Werner Heisenberg encountered this kind of problem with subatomic particles on the quantum screen. He discovered that if the particles were small enough, it was impossible to know their exact location and their momentum (mass x space/time) at the same time. If you know exactly where one of them is, you can know nothing about where it is going or how much mass it has. The more you know about its mass and velocity, the less you can possibly know about where it is. The Heisenberg Uncertainty Relation is as follows:

$$\Delta p \times \Delta x = h$$

where "Δp" is uncertainty in momentum, "Δx" uncertainty in space, and "h" the Planck constant again. Right off we can tell that we are dealing with extremely small masses, distances, and times, and that the quantum screen is so finely grained that it cannot be detected on the macroscopic level. It takes objects as small as protons and pi-mesons to make fuzzy patterns; there is no wonder it was not noticed until 1900. But the screen is not perfect; as in the case of relativity, there is a small but unnoticeable "quantum effect" even at macroscopic dimensions. When objects or parts of objects are between quanta, we cannot say where they are, or if they are at all.

It is important to emphasize here that we are not talking about any technical limitations in our microscopes or cloud chambers or particle accelerators; we are talking about a limitation in the world itself. Like the lights on Pac Man's video screen, "points" in space-time-mass are not quite touching one another. No matter how good our instruments, we will never be able to know an object's momentum and location precisely because the dimensional structure within which we perceive images is imprecise. There is nothing wrong with our information, it is the medium through which we are aware of it that is imperfect.

Tunneling

Another way to say that the location of a particle is uncertain is to say that its location at any given point is only probable. If I do not know where my keys are, I could say that there is a probability that they are in the closet and another probability that I left them downstairs. I can even say which is more probable by quantifying each probability: 60% chance for the closet, 30% for downstairs, 8% for the garage, and 2% for farther out in the universe. We could interpret this probability relation as a sort of "cloud of keyness" or a "keyodynamic field" that spreads out through space in and around my house with local points of concentration in certain areas. When I find my keys, wherever they may have been, the probability relation becomes useless and the field collapses to zero. This is how we have to understand the "location" of subatomic particles: there are only probabilities for finding them at particular points. The difference is that, on the macroscopic level, I know that my keys are "somewhere" in space-time, where on the quantum level there is no "somewhere" between quanta.

Because of the uncertainty principle in quantum mechanics, physicists express the location of particles in space in terms of probability relations and fields. But keys can exist anywhere; it is subatomic particles that can exist only at points in space that are not quite touching one another. An electron, for instance, can only exist at discrete intervals of space from a nucleus, or at distinct "energy levels:" the farther from the nucleus the higher the energy level. When an electron jumps from one level to another the way Pac Man jumps from one set of lights to another, this is called a "quantum leap." As its popular usage implies, a quantum leap is a jump into a completely distinct set of circumstances; there is no gradual transition between one level and another. In fact there is nothing, not even space itself, between one level and another.

If my keys were subatomic particles, they could *only* exist at certain points, like in the closet, downstairs, the garage, etc. The probability of finding them would be high at these places and disappear at places in between. The keyodynamic field would be a wave, with a large crest at the closet and smaller crests downstairs, at the garage, and so on. When I go to find my keys, they are almost certainly at the wave crests and not on the stairs or on the dresser top just outside the closet door. This "quantized space" would be very difficult to get used to, but similar to what real space is like at microscopic dimensions according to many physicists.

Now if I were to get up one morning and find that macroscopic space had somehow become quantized overnight, and I began looking for my keys, I would be likely to find them right there in the closet. They would probably be there the next time I look, too. But remember that I could not know exactly where they were: they would not exist at a distinct point in space, but as "keyness" spread out in space like a cloud. Sometimes "keyness" would be downstairs, and every once in a while, in the garage or even down the street somewhere. I could never tell, at any particular time, where the keys would be: the same "keyness" would yield them in completely separate places at different times no matter where I saw them last. I may find them in the garage the next time I look. How would my keys be getting from one place to another? Would they be "tunneling" their way through the wall of the closet and float down the stairs?

The tunneling would be only apparent. The keys would not exist anywhere in between points in quantized space; in fact they would not exist at all except at the moment they were "found," or perceived. They do not, therefore, move through the wall or down the stairs when we are not looking. When we are not looking, there are no keys. Another way to understand this is in terms of object identity: I assume that the keys I see in the closet are the *same* keys I later see in the garage, and therefore that they had to move through the intervening space to get there. But all I really see is keys in the closet at one time and keys that look the same later on in the garage.

Like my keys in quantized space, a subatomic particle at times appears to "tunnel" through impassible barriers in space as it leaps from one quantum level to another. An alpha particle trying to escape from a nucleus, for instance, would be like a tennis ball trying to pass through a brick wall. The energy of the alpha particle itself is many millions of times smaller than the energy of the force holding it in the nucleus. But because the alpha particle is in quantized space, it has probable existences over a range that may slightly exceed that of the nuclear force. Almost all the peaks of its wave function lie within the nucleus, but even the slightest ripple of a wave outside the nucleus means that at some time or another the particle will "miraculously" escape, as if the tennis ball were suddenly to appear on the other side of the brick wall.

We know this could never happen for the tennis ball, because no matter how hard or how many times you throw it or how long you wait, it will never just pop through the wall. But it does happen for the particle. It is impossible to say when, but at some point the alpha particle will simply appear outside its nucleus.

But how can it go through something much stronger? And where is it and *what* is it when it is on its way? The concept of space as a context for the particle breaks down at this level because the particle is smaller than the points of the "screen" on which it appears. The quantum screen is too coarse for subatomic images.

The image that leaps between points on the quantum screen is just as real as any macroscopic image, but the concept of "object in space-time" reaches its limits. We are dealing here not with a speck of matter, but with a speck of information; the problem is in the context for that information. Light information consists of tiny points of touch in a dimensional pattern. Dimensions break down on the quantum level because the sensory potentials corresponding to dimensions become indistinct. "Seeing" becomes "touching." Space-time melts into mass as visual consciousness melts into tactile consciousness. Because we cannot see anything smaller than a photon, the structure of space-time that we derive from patterns of photons does not work for objects smaller than photons. The image of a subatomic particle, therefore, appears to tunnel through the non-space between quanta.

Because an image is more fundamental than its dimensional context, it appears on the quantum screen as a "probability cloud" spread out through space. Subatomic particles are "clouds" that are often smaller, in five dimensions, than the quanta of which the screen consists. We cannot say that they exist at any particular quantum on the screen, only that there are probabilities for their existence at particular quanta. Subatomic particles in space-time-mass are like a Pac Man smaller than the little lights on his video screen. He could pop up anywhere; we can only say that there is a chance that he may light up a particular point.

The Two-Slit Experiment

When you throw a stone into a pool of water, circular waves radiate out in all directions from the point at which the stone entered the water. If you throw two stones into different parts of the pool at the same time, waves from each radiate outward and interfere with each other. Where one wave meets the other, they "add up:" the interference pattern will have wave crests as high as both original waves put together, and troughs as low as both together. At points where crests from one wave meet troughs from the other, they cancel each other out.

The same thing happens with waves of light. We can simulate the two stones in water by cutting two tiny slits in a screen in front of a light source. On the other side of the slit screen will be another screen without slits. Monochromatic light waves from the source will go through each slit in the first screen and radiate outward, interfering with light waves going through the other slit. The resulting interference pattern will show up on the second screen as alternating bands of light and dark. If one of the slits is closed, light waves no longer interfere with one another, and the alternating bands disappear. Open it again and the bands reappear. This experiment was used for many years to prove the wave nature of light.

But what if we drastically reduce the intensity of the light source so that only one light particle (photon) passes through the slit screen at a time? It must go through one slit or the other, and as it goes through, there will be no other photons around for it to interfere with. We will put a photographic plate on the second screen and allow the photons to accumulate over time to see if they still make bands indicating wave interference. To our surprise, they do! Each photon, as it passes through one slit or the other, seems to interfere with itself! If we close one slit, the bands disappear, indicating no interference. But if the photon goes through only one slit, how does it "know" if the other slit is open or closed? How can it determine whether or not to make interference patterns? Does a single photon, perhaps, go through *both* holes?

In a way, it does. Light behaves like a particle sometimes and like waves at other times, but never like both at the same time. When a photon hits the photographic plate and disappears, it makes a dot at a specific point in space, acting in this case like a particle. But in its flight "through space" between the source and the photographic plate, it behaves like a wave, even if reduced to individual photons. As a wave, the photon is like a cloud spread out through a small region of space; the particle is not at any particular location within the cloud, only more likely to be found where the cloud is thicker. Part of the cloud goes through one slit and part through the other. The cloud takes on a new shape as it emerges from the slits, and this affects where the photon will strike the photographic plate. We cannot say exactly where it will strike; all that can be known from the cloud is the probability of its striking a particular point. Over time, however, with many millions of photons following the law of averages, the interference pattern emerges as alternating bands of light and dark.

But here again, we are talking about photons as if they were little balls of light flying through space and landing on things. We have been speaking in terms of photons "from" the source, "passing through" slits, "and then" making a dot on a photographic plate. This, I believe, is the problem. If photons, instead of being particles in space-time, are tiny points of tactile sensation arranged into patterns of visual information, we are looking in this experiment at the basic context within which those patterns are arranged, or at what I have called the "quantum screen."

The two-slit experiment is a means by which the coarseness of the quantum screen is magnified to macroscopic dimensions. By allowing light waves to interfere with one another, we are "prying open" a space-like gap between quanta and projecting it macroscopically in terms of dark bands onto the photographic plate. By observing only one photon at a time, we are "prying open" a time-like gap, looking at the screen "edge on" and watching each individual light quantum make a dot at a specific point on the plate. If we allow dots to accumulate over time, the time-like gap re-closes and the space-like gap reappears in terms of the dark and light bands.

I will explain in more detail what I mean by the quantum screen and this interpretation of light in the next chapter. But before that, I will say something more about what modern physics has revealed about patterns on the screen, or what we call physical objects.

Matter Waves

In 1924, a French physicist by the name of Louis de Broglie suggested that if light had particle-like characteristics, matter might have wave-like characteristics. He reasoned that if the wavelength "λ" of a particle of light is equal to the Planck constant "h" divided by its momentum "p:"

$\lambda = h/p,$

then the "wavelength" of *any* particle should be equal to the Planck constant divided by its momentum. Again, we are talking about a phenomenon of extremely small proportions that exists at all dimensions but would be measurable only on the subatomic level. (The "wavelength" of our baseball discussed above would be on the order of 10^{-34} meters, or several quintillion times smaller than the diameter of a proton. The uncertainty in the exact location of the baseball, as represented by this wavelength, would be unlikely to affect the umpire's call from behind the plate.)

The difference, according to de Broglie, between a "material" particle and a light particle is that the momentum of a material particle is resolvable into separate quantities of "rest mass" and velocity. Photons have momentum, like macroscopic objects, but no rest mass. (They can have no mass because they accelerate instantaneously to "c.") The "p" of photons cannot be broken up into mass and velocity because mass is always zero and velocity always "c." Other particles, however, have momentum equal to rest mass multiplied by velocity:

$p = mu$

where "m" is mass, and "u" velocity. The faster an object goes or the more it weighs, the greater it's momentum. $\lambda = h/p$ for the photon becomes:

$\lambda = h/mu$

for the particle.

At the time, this was no more than an educated guess on de Broglie's part, but it was shown later to be true by an experiment similar to the two-slit experiment described above. Instead of a light, a beam of electrons was used, as in *Figure 3*.

If we close one of the holes (*Figure 4*), the electrons form a single "lump" on the photographic plate behind the screen.

If we open the other hole, we might expect that the electrons would simply form two separated "lumps," one behind each hole, as in *Figure 5*. This would be the case if electrons were like macroscopic objects and had no measurable de Broglie waves.

But when the experiment is performed, electrons do not behave as in *Figure 5*, but form interference patterns as in *Figure 6*. This means that they have the same wave-like characteristics as photons, thus proving de Broglie's theory. "Matter" behaves just the way light does, the only difference being its mass.

The electron's wavelength is more difficult to detect than the photon's because of the electron's value in the mass dimension. This increases its momentum, and thus decreases its de Broglie wave.

If visual images are patterns of photons (and physical objects patterns of quanta), as I have suggested, the mass dimension arises as part of the pattern. It is a relation amongst quanta, a parameter of "wholeness" to the pattern that is more than the sum of its parts. On the macroscopic level, an object's mass, along with its length or width, maintains its identity through time; it "holds" the pattern together against a background of potential randomness. On the quantum level, the mass of a subatomic particle such as an electron becomes indistinguishable from space and time, blending into the quantum screen in the form of de Broglie waves. The smaller the mass of the object, the smaller its momentum and the greater its wavelength. The greater its wavelength, the more it spreads out in space and time across the screen: a subatomic Pac Man with a new dimension.

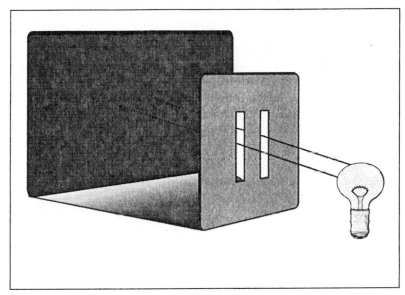

Figure 3

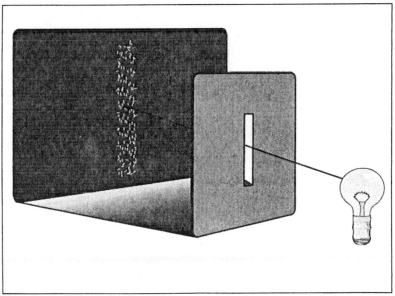

Figure 4

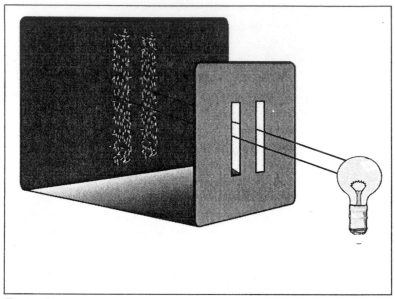

Figure 5

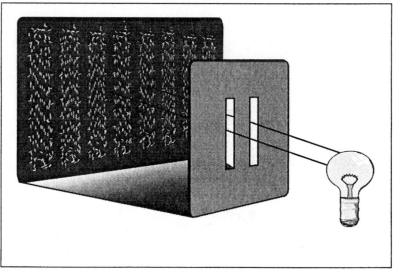

Figure 6

The Mach Principle

There is another enigma important to this theory that is found neither within relativity theory nor quantum mechanics. The phenomenon of inertia is a very familiar one, found in everyday life, and well understood from an operational standpoint. Its cause, however, remains a mystery. The problem seems to be that a sort of absolute space is necessary to account for it. How can a massive body "know" that it is being accelerated unless it is accelerated in relation to something?

Inertia was not a problem in classical physics because absolute space (as defined by the ether) was generally assumed to exist. Every object in the universe, including the Earth, the sun, and each star occupied a specific spot in the universe at any given time, and tended to stay at this point or move uniformly through it. Any tendency away from an existing position in space or straight-line velocity through space was resisted, and could be "felt" by a body. It is for this reason that stationary bodies tend to stay at rest and moving bodies tend to remain in straight-line motion. The resistance was caused by space itself. But for Einstein, and for others who rejected the idea of absolute space, inertia created serious problems. What, exactly, was an accelerating body accelerating against?

An operational solution to the problem was suggested by Ernst Mach and accepted, at least temporarily, by Einstein. Mach proposed that the "fixed stars" posed a sort of rigid system that every massive body could somehow sense. The total mass of every object in the universe, no matter how distant, produced a sort of "average" space which influenced the motion of each individual object. Mach did not embarrass himself with any suggestion as to how this was possible, and continually emphasized the operational nature of the idea. If there is no absolute empty space, it must be the objects *in* space that form a reference frame for inertia. It is amazing that a skeptic like Mach, who did not even believe in atoms because he could not see them, would come up with so flimsy a speculation.

Another problem associated with inertia is the kinesthetic sensation associated with the acceleration of an observer. When an observer is accelerated, he reports a uniform kinesthetic sensation throughout all parts of his body in proportion to the second time dimension of his motion. This "g force" is normally explained as due to cell compaction within the observer's body. But there is nothing to account for cell compaction other than absolute space or the Mach Principle! Because this sensation is uniform (it is not felt in any one part of the body or just on the surface, and involves the entire tactile realm of consciousness) it can be directly identified with the entirety of the second time dimension. When an observer feels acceleration, he experiences a dimensional shift in relation to the whole of four-dimensional space-time, or a "curvature" of four dimensions into five. This is the same "curvature" that a stationary observer, according to Einstein's principle of equivalence,[17] feels in a gravitational field.

The failure of classical and modern physics to explain so common a phenomenon as inertia is another indication that the fundamental assumptions upholding modern science are inadequate.

The enigmas mentioned here demonstrate that dimensions and dimensional concepts lose their meaning at dimensional extremes. Perceptual images that seem "in" the world on the macroscopic level cannot be contained on the quantum level nor at extremely high velocities by space, time, or mass. The world that we know in everyday life exists only at the middle latitudes of these dimensions and disappears at the fringes.

In fact, the everyday world of space, time, and mass does not exist at all except as a background for perceptual images. Where sensory information is separable into distinct realms, dimensions exist; where they merge, dimensions do not exist. What we call the world, then, is not something that we actually experience, but a background against which the information that we do experience becomes perceptual.

VI

LIGHT

A s we have seen in relativity theory, light imposes limits on space-time, setting an absolute limit for the velocity of any object and distorting the dimensional relations of objects that approach that velocity. But what does light have to do with how fast something goes? If light is a thing in space-time, how can it affect the length, the mass, and the passage of time of other things in space-time? And why should the equivalence of mass and energy have anything to do with the speed, much less the *square* of the speed, of light? The enigmatic behavior of light and its effects on "material" objects are well known, but there has never been a satisfactory explanation of what light is, and how it is capable of so much. We seem to understand light physically but not metaphysically.

Light as a Tactile Medium

The enigmatic properties of light are due, I believe, to the fact that visual information is reducible to tactile information. Photons are tiny tactile points that we "feel" and dimensionally-ordered patterns that we "see." A photon by itself is fundamentally non-visual, but photons in the aggregate constitute the visual realm of consciousness. Vision breaks down at the level of individual photons, as a single photon cannot be said to constitute a pattern. As the visual potential blends into the tactile, space-time and mass become indistinguishable.

It is this "tactile reduction" of light that is behind the dual nature of light: the fact that light can be understood as waves or as particles, but not both at the same time.

The tactile reduction of light does not, however, mean that light is nothing but tactile sensation in a different form. Patterns of photons are not the photons themselves, but the order in which they occur. It would be absurd to say that a building is "nothing but" the bricks of which it is made. The building, as a particular pattern in which the bricks occur, constitutes a wholeness; there is a distinct reality to the "information" supplied by the bricks over and above the bricks themselves. A visual pattern, therefore, is a whole that is more than the sum of its tactile parts.

Photon patterns become quantum patterns that we can touch, hear, taste, and smell because they are information located in potentials corresponding to these realms. But the disintegration of these sensory potentials on the quantum level means that what we experience on that level are not physical objects as such. Protons, quarks, and pi mesons are not objects. They are every bit as real as tables and chairs, but they do not fit into the dimensional structure of perceptual consciousness as do macroscopic objects. Like water through a sieve, they slip through the macroscopic net in which we catch the perceptual patterns of everyday life.

It is interesting that the structure of perceptual consciousness is least apparent at the point where it takes effect. We are so used to the existence of physical objects in dimensions on the macroscopic level that we do not notice sensory potentials or think of them as a structure of consciousness. The habit of sorting sensory information into separate realms is so fundamental to our world view, and so deeply ingrained, that it has long since become unconscious. It is only with recent experience at the extreme edges of perceptual consciousness that the fabric of space-time has become unraveled and the fibers revealed.

But what are those fibers, and exactly how are they interwoven? The fibers are space, time, and mass. They are *created* at the quantum level and interwoven in terms of the universal constant "c," and its square, "c^2." "c" expresses the fundamental unity in nature between space and time: one second is equivalent to "c" meters. It is not a velocity. It is misunderstood as a velocity because it is expressed in the same dimensional terms: "meters per second." A velocity is a relation between values within two distinct dimensions. There is an upper limit to velocity because its dimensional components are not ultimately distinct. As a relational structure of space-time, "c" is also a limitation of space-time.

The "c" limitation is the impossibility of separating space from the time of any event. One second is the same thing as 10^8 meters. An event observed 10^8 meters distant is always one second old. This is apparent in astronomical observations at extreme distances: it is impossible (from a single observation) to separate the distance of a visual image from its age. A star's image at ten light-years is always ten years old, and at one hundred light-years, one hundred years old. Light is inherently limited in this manner as a medium of information. The same limitation exists to a very small extent in everyday life: even light crossing the room is a little bit "old." As we look at objects at various distances around us, we are also looking, however slightly, into the past. That we do not notice this limitation resulting from the fundamental equivalence of space and time is due to the fact that "c" is a very large number.

But what about mass? If space and time are fundamentally equivalent according to the relation "c," how are they related to mass? This gets back to how we actually experience light.

The tactile realm, we remember, in the absence of seeing, hearing, and smelling, is experienced without space dimensions. An observer who experiences only tactile sensation can have no sense of "world" apart from "body." As a tactile sensation, therefore, a photon carries with it a sense of time but not of space. As a point of *light*, however, a photon is a vibration in five dimensions. Space is *created* as vision is created. The photon is experienced in new dimensions structurally based on the relation "c." Its mass value is expanded to a five-dimensional value, energy, when the photon is perceived in the visual realm. This is space-time-mass. Einstein's famous equation:

$$E = mc^2$$

demonstrates the structural relation between tactile and visual perception. Vision arises as a separate realm of perception as minute tactile sensations are arranged into information in a dimensional context.

Einstein's equation states that the energy of a photon is equal to its mass[18] times the square of the structural relation between space and time, or that a unit of visual consciousness is equal to its tactile value expanded by four dimensions. Put another way, any value in the mass dimension can be either touch or vision; if this value is vision, it will be experienced in the form of trillions of photons of energy, "mc^2." Actual vision, therefore, is composed of a very small amount of "touch."

The means by which the touch experienced by a single cell becomes the vision of many cells is discussed in the next section.

Light as an Intercellular Medium

The development of space dimensions follows the rise of multicellular consciousness and the evolution of specialized sensory organs. Information passing between sensory organs and other parts of the body must be in a form that is at once understandable to the organism as a whole and reducible to a form that can be experienced by a single cell. Seeing and hearing in multicellular organisms, for instance, consist of the compounded experience of individual cells in the retina and tympanum respectively. Single cells in the retina or tympanum, however, do not "see" or "hear" in the normal sense; what they experience individually is limited to chemical or tactile stimuli. Other cells in the organism do not experience these stimuli directly, but indirectly through coded stimuli in the form of information. The means by which information is generated by sensory cells and received by cells in the rest of the organism is what we will call an "intercellular medium." Light is an intercellular medium in that it carries information from one cell to another the way that an electronic medium carries information from one person to another.

We normally think of vision or hearing as an experience of the organism as a whole, rather than of individual cells within the organism. But what, exactly, is the organism "as a whole?" Is its "being" something in only some of its cells, in all of them collectively, or somewhere in between individual cells? At what point, in evolutionary terms, does the wholeness of a simple, multicellular organism such as a sponge, supercede the individuality of its separate cells? Obviously, the wholeness of the organism cannot be reduced to a location within individual cells, but depends upon collective functioning. It also depends upon collective experience. The more uniform the experience throughout the cells of the organism, the more highly defined its wholeness. A sponge, having no central nervous system or sensory organs, has very limited uniformity of experience among its cells, and therefore a limited wholeness. More highly developed animals with eyes and ears, however, encode information from these organs in a form that is so uniformly experienced throughout the organism that individual cells lose their separate identities. Intercellular media, then, are not only the means by which sensory cells communicate with other cells, but constitute the organism itself as a whole. The "being" of an animal is the uniformity of its cellular experience.

The connection between cellular experience and that of the organism as a whole is what I have called the "tactile reduction" of sensory information: individual cells experience the "touch" of photons while the organism as a whole "sees." What we call the quantum level of the physical world is multicellular experience reduced to the experience of individual cells.

Light, sound, smell, taste, and the body are all intercellular media. They consist of potentials intercoordinated in such a way as to produce what is experienced as the dimensional world. An intercellular medium is the same thing as a dimensional potential seen from a different point of view: one is biological and the other physical. From the physical point of view we have seen that macroscopic consciousness consists of composite patterns in a dimensional context, and from the biological perspective that it consists of composite cellular experiences within the context of an organism. As a context for sensory information, a medium and a potential can be understood to be more or less synonymous.

But how does actual sensory information in one realm become "potential" information in the others? This has to do with the relation of each medium to the others. If sound and light are each separate media, and thus separate potentials, how is it that they are coordinated into a single world? We have seen already that as sensory potentials become dimensions, they develop an *orthogonal*, or right angle relation to one another. In this way, distinct forms of information interrelate without interfering. A visual image never shows up in the auditory realm, but indicates the location of a potential auditory image. We are able, in this manner, to see and hear objects in a single, multidimensional world without seeing sounds and hearing sights.

The orthogonal relationship of light and sound can be demonstrated physically by the following thought experiment:

Two observers moving uniformly relative to one another are trying to determine which is at rest and which in motion. Each is emitting sound and light waves from objects in his immediate vicinity and in his own frame of reference. Each will report an auditory Doppler shift as the other passes by, but one of them (Observer A in *Figure 7* below) will report a considerably greater shift than the other. This observer concludes that he is at rest in relation to the sound medium and the other in motion. This is because the wave crests from a stationary object form spherical shapes around the sound source, while wave crests from an object in motion form elliptical shapes. The greater doppler shift he reports is due to the elliptical shape of the sound waves coming from the other observer. The observer in motion (Observer B) agrees with this analysis, and discovers that he could determine the direction and magnitude of his motion relative to sound without consulting the stationary observer, merely by moving his sound source to different locations around him and measuring the wave ellipse through differences in pitch. Sound waves in the direction of his motion are closer together and of higher pitch than those in the opposite direction. Also, if he is moving faster than sound itself, he will report sounds from the stationary observer, but the stationary observer will not be able to report any sounds from him as he moves closer. The stationary observer will, however, report a sonic boom as the moving observer passes by. The moving observer will not report the boom. There is, therefore, an absolute space-time relation of each observer to the sound medium.

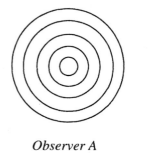   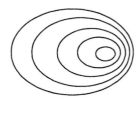

Observer A   *Observer B*

Figure

If each observer examines the light from his own and from the other observer's light sources, he will be able to detect no such

absolute position in space-time. There will be no way to determine
which is moving and which is not. For each observer, light waves form
spherical shapes as they move off into space from his own source and
elliptical shapes as they move off from the other's source. (Their relative velocity must be a significant fraction of "c" for this to be noticeable.) Each observer will think *he* is observer A and the other is
observer B. This is revealed as each observer detects no difference in
color (corresponding to pitch differences in sound) as he moves his
own light source around him, but reports a color shift as the other
passes by. But the color shifts reported by each are the same. (Each
sees the same color shift, or the same elliptical shapes, in the *other's*
light waves.) There is, therefore, no way to determine an absolute relation of either observer to the light medium.

The sound medium can be detected in space-time, therefore, where
the light medium cannot. But if light consists of minute tactile sensations, and it is the second time dimension (mass) that constitutes the tactile potential, perhaps light can be detected in space-time-*mass*. Perhaps
we are looking for the "luminiferous ether" in the wrong dimension.

To test for this, we must not only move one of our observers, but
accelerate him, and see if this constitutes absolute motion in relation to
light. He will report two related phenomena: first, (if his acceleration is
great enough) he can detect *elliptical*, rather than spherical, light waves
coming from his *own* light source; (the same shape as the sound waves
from his own source in uniform or *non*-accelerated motion). From this
he can tell that it is he that is accelerating, not the other observer, and
that his motion is absolute in relation to light. And second, he reports a
uniform "g" force proportional to his acceleration; in other words, he
"feels" the medium within which light is experienced.

He can measure his absolute motion in relation to light by measuring the shape of the ellipse of light waves around him. He may do this
by measuring differences in the value of "c" in various directions. He
will measure "c" to be less in the direction of his acceleration and
greater in the opposite direction. (Time will be worth fewer meters
ahead of him and more behind him.) This structural shift in the value
of "c" is, for the accelerating observer, a dimensional interchange
between mass and a second space or time dimension.

The accelerating observer knows that he is Observer B, and that the
other is Observer A. But the significance here is that we may use the

same diagram we used to describe the non-accelerated motion between observers of sound. If we are granted that there is an orthogonal dimensional relation between acceleration and constant velocity, we may conclude, by the congruent elliptical shape of sound waves from a source in constant velocity and of light waves from a source in constant acceleration, that the medium of light is orthogonal to that of sound.

We may also conclude that the medium of light is the tactile realm, and that to detect it we must move in the dimension to which it corresponds, the second time dimension.

Light as a Universal Medium

Each medium thus adopts a single universal structure and can be inter-coordinated with every other medium. In an orthogonal (right angle) arrangement, every point in one medium is also a point in the other four. A photon, for instance, is a point of light, but it is also a point in four other dimensions of space-time in which there are potential perceptions. Each realm remains distinct from the others, despite intersections between all realms in the form of physical objects.

But what "universal structure" is adopted by all five intercellular media? The answer, as we have already seen, is the conversion of four of them into the fifth; light becomes a universal medium for all realms. The fundamental "bit" of visual consciousness, the photon, is abstracted into a "potential photon," or quantum, and becomes a point in space-time-mass, the context of all macroscopic perceptual experience. Each medium except light thus becomes an abstraction experienced in terms of light. Objects heard, smelled, felt, or tasted are "envisioned" as ranges of potential photons, whether or not they are actually seen. Space and time, though derived from light, are extended beyond actual vision to encompass all of perception. A single multidimensional "quantum screen" becomes the entropic background for perceptual consciousness.

This is why a medium for light cannot be detected in space-time. We define space in terms of light (a straight line in space is defined as the path of a beam of light). There is thus no way to find light in space.

Does this make a visual image, or pattern of photons, the same thing as the object itself? Experimental confirmation of special relativity indicates that what is experienced in terms of light is true of space-time itself: space really does shorten and time really does slow. The photons really are the object. Furthermore, the effects of relativity remain whether or not they are experienced visually. The object acts like a pattern of photons even when there are no photons: the space-time structure of the object, derived from light, remains even without light. If the object is experienced visually, the photons return. The pattern, therefore, consists not only of actual but of *potential* photons.

But why is it that the structure of space-time-mass is derived from light and not from some other realm? Would it not be as convenient to structure perceptual consciousness from the dimensional relations found in sound? At sea level, sound travels at the rate of 331 meters per second. Why is it that one second is not equivalent to 331 meters in space-time?

Most likely, it is because "c" is so much larger than 331. If space and time were derived from sound, dimensional extremes would be much closer to macroscopic experience, and relativistic and quantum enigmas that much closer to everyday life. Air molecules would compose too "coarse" a medium for consciousness as a whole. Photons are much smaller, in terms of energy, and therefore make better quanta of perception. Also, because air molecules are so much larger than photons, each can be understood as composed of many quanta, and the medium of sound thus incorporated into a light-based space-time. Sound "fits into" light much better than light into sound. The medium of sound can be detected within space-time because sound "quanta" can be understood in terms of light quanta.

A space-time based on sound could serve as a structural basis for chemical, tactile, olfactory, and auditory consciousness,[19] and may even do so at certain evolutionary stages. Space-time derived from light may not have served as the dimensional groundwork for perception at all times, as the visual realm of perception has not existed at all stages of evolutionary development.

The fact that perceptual consciousness as a whole is structured on the dimensional relations of light means that information in the non visual realms is interpreted in the same way as light is interpreted. An object is heard or touched where it is potentially seen. Sounds heard in the distance or objects felt out of the field of vision are "envisioned" as images in space-time-mass based on the fundamental relations "c" and

"c^2." Potential photons are points in space-time-mass; physical objects are patterns of photons, potential or actual. The graininess of space-time-mass is due to the fact that these points are not infinitely small.

We are ready now for a model of perceptual consciousness as a whole.

The Photon Screen and the Quantum Screen

The quantum screen that we have been discussing is an extrapolation of what we may call the "photon screen" and serves as the context for all perceptual experience. To distinguish the quantum from the photon screen, one should understand that the photon screen is the visual realm only; it is light itself as experienced by an observer, in the absense of any other realm. The photon screen is the four-dimensional background against which changing points of colored light become orderly events and objects. It is similar to a computer or television screen with an added space dimension. The screen is not in space or time; it *is* space and time. Points on the screen are extremely small, but not infinitely small. Visual images, therefore, are of very high resolution and appear perfectly smooth on the macroscopic level. It is only the smallest objects, those smaller than the points on the screen itself, which appear "fuzzy."

Each point of light is a photon of a specific frequency at a specific point in three dimensions of space and one of time. Its frequency is expressed as a magnitude "per second", or as a value in an additional time dimension, the same dimension as mass. Each photon is, therefore, a point in three dimensions of space and two of time. The second time dimension of each photon is "felt" as mass and "seen" as color. But what is the relation here between mass and color, and how is it that they are experienced in the same context?

The context is the same only on the quantum level, where the color of an individual photon is experienced in terms of its tactile value, a value in the mass dimension. On the macroscopic level, where the visual and tactile realms are distinct, color is experienced as an actual perception on the photon screen, but mass is not. The photon screen is limited to four dimensions: the frequency, or "per second" value of each photon is experienced only as color, and not as a value within a dimension. We cannot "see" mass, at least not in a single visual image. Mass can only be constructed within visual consciousness through a series of time-separated images. We must "sniff out" the mass dimension of physical objects on the photon screen the way an olfactory observer "sniffs out" additional space dimensions. We must compare several visual images in order to perceive acceleration; first we must measure the velocity of an object by measuring the distance it covers between two points in time, and then measure its velocity again at a later time with at least one more distance measurement. Then we must compare the two velocities to determine acceleration, which is the only way to measure mass. Mass is in this way "foreshortened" in the time dimension of the photon screen, the way depth on a two-dimensional picture of a three-dimensional object appears foreshortened in the length or width dimensions.

But as we look very closely at the photon screen, we notice that it differs from a computer or television screen in that the points are separated slightly not only in space, but in time and mass. Extremely small visual images are fuzzy in space and time, and as they approach the size of photons, disappear into the mass dimension as well. The mass of a *pattern* of photons, which shows up on the screen as the "per second" of acceleration, disappears into the mass of individual photons as the "per second" of frequency: hence the de Broglie waves of subatomic particles. We find that the closer together points are in space, the farther apart in time and mass, and vice-versa.

If we look closer still, we notice that there is no physical manifestation of the screen itself; there is no screen in between points on the screen. It is merely a conceptual context in which vision is experienced. For macroscopic objects, the gap between points on the screen can be bridged easily, but when the "size" of objects approaches that of individual points, the screen no longer works as a proper context for perceptual experience.

The photon screen we have described here is only one of the five realms of perceptual consciousness. But it has a special significance to the other four. If there were a separate screen for each realm, with its own type of points, perceptual consciousness as a whole would be fragmented and confusing. There would be many worlds instead of a single, multidimensional world. All perceptual experience is consolidated into a single world by incorporating auditory, olfactory, chemical, and tactile information into a *potential* photon screen based on the structure of the actual photon screen. "Bits" of non visual information are superimposed upon the photon screen in the form of potential photons. They may or may not be seen. Each non visual bit is much "larger," in five dimensional terms, than photon bits; an air molecule, for instance, as a "bit" of auditory information, is much larger in terms of energy than a photon. It consists, therefore, of a pattern of many potential photons.

But how are potential photons experienced if they are not actualized? Do we *hear* an object in the next room somewhere in *visual* consciousness? In a way, we do. The screen, originally a context for the interpretation of photon information, is extrapolated in all directions beyond vision to include each of the other realms. When we hear a voice in the next room, or smell the neighbor's barbecue, we "envision" physical objects in a dimensional context that is potentially seen. When we touch something in the dark, we experience potential photons at specific points in space and time; when the lights go on, we see them.

But what are potential photons? Are they what we actually touch and hear and taste? Yes, but only in the aggregate. There is no way to experience an individual potential photon except as an actual photon. Patterns of many potential photons are the form in which all perceptual information is experienced. They are points in five dimensions, or fundamental "bits" of perceptual consciousness, actual in any realm and potential in all realms. They are the quanta that Planck first discovered in 1900; the basic units of energy itself.[20] Perhaps the best way to understand them is as abstracted photons, stripped of their actual manifestation as points of light, left only with their dimensional location. They are actualized individually only in visual consciousness, when they "light up" on the screen. The "quantum screen," then, is perceptual consciousness as a whole.

As the space-time-mass we experience in perceptual consciousness is structured after the dimensional relations of light, there is a special relationship between the quantum and photon screens that we have described. The photon screen is what we see of the quantum screen, or the colors of individual photons that occupy the mass, or "per second" dimension of individual quanta. As we move our bodies, or just our eyes, we rotate dimensional axes and actualize new quanta in the form of light. It is interesting that the fundamental structure of perceptual consciousness, the quantum screen, is purely conceptual, but derived from a form of actual perception.

But how is it that we feel ourselves to be *on* the quantum screen, that is, "in" space-time? The tactile realm, we remember, existed without any space dimensions at all before the evolution of the olfactory, auditory, and visual realms. How did the body end up *in* space?

Tactile consciousness, in the absence of the higher, "spatial" realms, is not in the form of quanta. Tactile experience is only quantized in order to coordinate it with other types of experience. Without visual, auditory, and olfactory information, there is no need to experience tactile sensations in the context of a universal intercellular medium. The chemo-tactile world of single-celled or simple multicellular organisms is consequently without space dimensions; there is no actual experience anywhere but in the body, and no potential experience anywhere but in the body. The body and the world are the same thing; neither has any existence apart from the other. The world and the body become separate only with the quantization of tactile experience which occurs with the development of specialized sensory organs and corresponding intercellular media. At this point the body becomes actual tactile perception within the tactile potential. It is with the evolution of complex multicellular consciousness that coordination of tactile experience with spatial experience becomes necessary, and a universal medium such as light becomes necessary. To coordinate tactile with other types of sensations, they are "quantized," or experienced as information located in a dimensional potential based on the structure of light. This potential is, of course, the second time dimension. The relation of the body as a *whole*[21] with the second time dimension is experienced in the form of a "g" force every time the body is accelerated. The tactile realm, then, exists initially without any space dimensions, but once quantized, becomes the body in space-time-mass.

Even as complex multicellular beings, it is possible to experience the tactile realm in its unquantized form, that is, without space dimensions. This is because actual tactile sensations remain the same whether quantized or not; it is only their context that changes. With the eyes closed, and with as few auditory distractions as possible, it is possible to experience tactile sensations without envisioning the body in space. Rather than feeling parts of the body as potential patterns in the visual realm, one can feel them directly, as pure kinesthetic experience devoid of any dimensional context other than time. Pressure "on the fingers" or strain "in the leg" can be experienced in terms of the sensations themselves and not in terms of occurring "somewhere." Even the concepts "finger" and "leg" can be understood as they are felt, and not as they are potentially seen. It is difficult to maintain this type of awareness for long, as there is a natural tendency to slip back into quantized tactile experience as soon as there is a need to coordinate it with experience in other realms. But with some practice, it can be accomplished.

With some additional concentration, this state of awareness can be maintained even as the eyes are opened. To do this, however, visual patterns must be in no way coordinated with experience in the tactile realm; patterns on the photon screen must remain detached from potential touch, as if they were on a computer monitor or television screen. If this can be accomplished, two separate realms of consciousness will be experienced outside of space. Patterns *on the screen* remain dimensional, but the visual realm *as a whole* will be experienced outside of space. The screen itself remains without shape or form, and like the tactile realm from which it has evolved, outside of any context that might give it shape or form.

But what, then, is the relation between the two realms? They cannot exist alongside one another, as "alongside" implies a spatial context. Each realm can be experienced only as a separate, unconnected world. This is possible, but very difficult, as we are used to relating separate entities in terms of dimensional concepts.

It is also difficult to maintain because of the dimensional connection between patterns *within* the visual and tactile realms. As soon as we discern distinct patterns within the visual realm, they become coordinated with potential tactile experience: the photon screen gives rise to the quantum screen. Tactile consciousness becomes quantized in order to make coordination possible: a touch sensation takes place in a spatial context only because that is where there is potential vision. The body becomes a shape on the quantum screen to make one world of two separate realms of consciousness; that world is quantized in terms of the fundamental structure of photons because light has become the universal medium of intercellular consciousness.

We can experience the tactile realm in either its quantized or unquantized form because there is no difference in *actual* tactile experience; only in the context in which it is experienced. No new realm is created as the tactile realm is quantized, only a new potential.

The relationship between the tactile and visual realms is the basis of the physical world that we experience on both the quantum and macroscopic levels. But it is a relationship both complicated and convoluted. Where the visual realm arises on the quantum level from the tactile experience of individual cells, the tactile realm of the organism as a whole arises on the macroscopic level in a spatial context based on vision.

The quantum screen is our model of perceptual consciousness. It is the entropic background of dimensional experience, that is, of so-called "material" images. It is not, however, the context of thought, imagination, or other forms of subjective experience, nor of observational experience. The context of observational experience is an extrapolation of the quantum screen analogous in many ways to the extrapolation that produces the quantum from the photon screen. This I will explain in the next chapter. In the next section of this chapter, I will say a few things about what constitutes an "object" on the quantum screen.

Object Identity

An object is a range of points in five dimensions, or pattern of quanta on the screen discussed in the last section. But the points on the quantum screen are not quite touching each other in space, time, or mass. A degree of conceptual input is required on the quantum level to link the points into a distinct pattern. The pattern appears only when we "stand back" from the quanta themselves, the same way that the picture in a newspaper photograph only appears when we stand back far enough from the ink dots of which it is composed. The pattern is the object; object identity, therefore, is a concept based on how close together the object's parts may be. The closer they are, the greater the degree of identity the object may be said to have.

This same principle operates on the macroscopic level. If an object's parts are "touching," there is a high degree of identity in space. But that identity can be lost in time if the parts tend to drift apart; "when" the object ceases to exist is a matter of subjective judgment. Object identity can also be established in space if the parts are close but not touching, the classic example being the forest and its trees. The closer the trees, the easier it is to see the forest. But where the trees thin out and the prairie begins, it is difficult to say at what point the forest ends. At the first blade of grass, at the last tree, or at some specific point in between? Like a pattern on the quantum screen, the identity of the forest is a concept that cannot be entirely contained in the bits of information by which it is known.

An object's identity is also established by the degree to which its parts constitute a common center of mass. The greater the force holding them together, the more likely the parts are to retain object identity through time. If the pattern stays together despite bouncing around against other patterns, it maintains the same center of mass. If the center of mass shifts slightly, the pattern changes its shape to some extent. Whether it remains "the same" object or not depends on a judgment as to how great the shift.

Without new energy, quantum patterns move on the screen with constant velocity only, and do not change their center of mass, thus retaining their identity. They remain "flat" in the second time dimension, never stopping, accelerating, or changing direction. But if two patterns collide, their centers of mass shift to some extent; their identities change momentarily as they decelerate, or shift in the second time dimension. Each pattern becomes distorted, as both cannot occupy the same points in space-time. "Room" to absorb the energy of collision is found in the second time dimension. If the patterns rebound quickly and object identities are restored, the collision is said to be "elastic." There is no permanent change in the identity of either object. But if one or the other object changes identity, the collision is "inelastic." Something permanent happens to the shape of the object; it may lengthen, shorten, gain or lose mass, or cease to exist.

In an elastic collision, each pattern is compressed briefly but then quickly regains its former shape. The kinetic energy of each (its energy of motion) may change as a result of the collision, but the kinetic energy of the two together does not. (Kinetic energy lost by one is gained by the other.) Each pattern retains its identity, though temporarily distorted during the collision.

In an inelastic collision, total kinetic energy is reduced and there is a change in the identity of at least one of the objects. If a ball collides with a pillow, the ball retains its identity, but the pillow comes away pushed in on one side. The loss of kinetic energy has to do with the new identity of the pillow; some of the energy of the collision, rather than going into the motion of the pillow as a whole, went into the motion of parts of the pillow in relation to each other, rearranging them and increasing their heat content. Total energy (energy of motion plus heat energy) is, of course, conserved. The new shape of the pillow will have a new center of mass.

Total energy remains constant before and after the collision, but the form of the energy changes. Losses in the energy of the motion of the object as a whole are made up with gains in motion of the parts of the object.

The concept of heat as random molecular motion can be turned around to make pattern motion a form of organized heat. An object at rest can be understood as molecules in random motion within a particular pattern; vibrations take place in all directions at once, each direction exactly balanced out by the rest. The same object in motion is a shift in the equilibrium of molecular motion in one direction or another. The average velocity of molecules is no longer zero, and the pattern moves. Each molecule moves in unison, to some extent, in the direction of the pattern's motion; molecular motion is no longer random.

If two such patterns collide elastically, they move momentarily in the mass dimension, and then back to where they were originally. Total molecular motion is the same before and after, and the patterns are the same before and after. But in an inelastic collision, the center of mass, or the average molecular motion, of one or both patterns shifts permanently. Total energy, or overall molecular motion remains the same, but the equilibrium of their motion has shifted. The loss in kinetic energy is a loss in the "organization" of heat. It is exactly compensated by an increase in random heat.

Gravity is a great complicator of object identity. It would be far easier to establish separate identities for each object if we were in interstellar space, far from any gravitational influence. We could easily determine where and when each object began and ended, and not have to worry about such things as friction. We could also more easily determine relative locations in the mass dimension as we watched objects bouncing off one another. But in a gravitational field such as that of the Earth, we have a tilt of dimensional axes into mass: all objects are accelerated uniformly at the same rate. As objects fall, they contact stationary objects beneath them that break the fall, but by virtue of that contact, become involved in the identity of supporting objects. Ultimately, all stationary objects in the Earth's gravitational field are part of the identity of the Earth itself; each is in contact with the Earth, contributes to the total mass of the Earth, and affects the Earth's shape and center of gravity to some extent. In fact, objects have no complete separate identity until they move relative to the Earth, and even then, are dragged gradually back to an Earth identity through friction.

The Earth's identity with objects in its gravitational field is never complete, however. A house or a mountain or a tree always retains a degree of separateness, even if it cannot be separated completely. The Earth is thus more of a "system" of objects than a single object.

All objects, on Earth and elsewhere, have dimensional relations to other objects, forming "systems" of greater or lesser identity. The closer the objects in space-time-mass, the greater the identity of the system. An automobile, for instance, is a system of interrelated objects like tires, fenders, carburetors, and steering wheels. If the objects are separated, the identity of the automobile disappears. Objects are to the system, then, what parts are to the object. In fact, seen from a distance, an automobile's constituent objects may not be apparent at all. A system is really an object seen up close, and an object a system seen from a distance.

This may seem intuitively obvious, but it is an important point that science has failed to recognize in its fruitless search for the "basic building blocks" of matter. Atoms were at one time thought to be the fundamental particles of matter. Everything was made of atoms, and only atoms. But atoms turned out not to be discrete particles, but systems of *sub*-atomic particles. These, in turn, were found to be not discrete pieces of any "thing," but systems of "something." The fact that it is impossible to establish distinct object identities for subatomic particles is due to the fact that there are no fundamental "things" that, when stuck together, form them. There are no "things" on the quantum level due to the failure of the dimensional parameters by which object identity is established.

VII

OBSERVATIONAL CONSCIOUSNESS

"You and I"

Everything we have discussed to this point takes place in a single perceptual consciousness: yours *or* mine. There is only one. Your consciousness is not in mine, nor mine in yours. We have removed consciousness from a dimensional context, and thereby removed the possibility for relating one consciousness to another. If it does not exist in space and time, where would another consciousness be?

But if there is no more than one, whose consciousness do we mean by "perceptual," yours or mine? One answer is to declare that if there can be only one consciousness, it must be mine. I have no way of knowing that "you" really exist at all. You move around and talk and jump back when you touch a hot iron, the same sorts of things I do, but this does not mean that you see and hear and feel; it only means that you react as if you do. You could be a highly sophisticated piece of computer-driven machinery programmed to respond to stimuli in such a way as to appear alive. You and the rest of the world could be no more than an idea in my mind; all I know is what I actually perceive. This, of course, is the solipsism.

The solipsism will never be a popular idea, however, as no more than one person can hold it at a time. Certainly, it would be difficult to get a room full of people to agree that the solipsism is the only way to understand modern physics and that only one of them was really there. Who gets to be the lucky one? What would everyone else do with themselves, and how would they understand what seems to be "their" consciousness? And how would they take attendance at the next meeting?

The solipsism has other problems, too. For one thing, it is morally repugnant. We have been taught from an early age to see things as others see them, to feel what others feel, and to do to others as we would have them do to us. But if I am the only real consciousness, there is no reason to care what happens to you or to anyone else. I should be free to use you as I would a tool or any other inanimate object. Once I fully understand the implications of quantum mechanics and relativity theory, I should reject moral behavior as unscientific, and act in as selfish and infantile a manner as possible. Egotistic behavior is ultimately logical, after all, in a world of pure self.

A good solipsist will tell you that it need not be so bad. I can best use you by being nice to you in order to get what I want. I can defer self-interest, temporarily, and thereby create a more civil order of behavior. The good moralist, however, would never accept a civilization, no matter how orderly, based entirely on deferred self-interest. Most of us would feel more comfortable with the moralist than the solipsist, but which of the two has a more scientific worldview? If we show, through the best and simplest interpretation of scientific evidence, that there can be no more than one consciousness, are we forced, ultimately, to reject morality in favor of the solipsism?

I think the solipsism is a good logical argument. It makes a lot of sense even though it is difficult to come to terms with. No matter how distasteful it may be nor how we may wish to fight it, the solipsism makes modern physics more understandable than does the "common sense" view of many consciousnesses in space-time, *at least for perceptual consciousness*. If it were the only alternative to the common sense view, it would be accepted, slowly and painfully, the same way that a curved Earth or the heliocentric solar system were accepted slowly and painfully. But where the solipsism fails is in dealing with the problem of communication between observers; it can explain what I see and feel, but it cannot explain why, through how you act and what you say, you appear to me to "have" consciousness. The solipsism can try to explain you as an extremely complicated feedback mechanism that reacts to stimuli and informs me of the existence of objects I cannot perceive directly, but it cannot do so gracefully. I reject it ultimately not because it is absurd or amoral, but because it is complicated.

I think there is a simpler way to explain the apparent existence of other consciousnesses (observers), that is consistent with what I have said so far about physics. It is a way that is not only consistent with the dimensional structure of consciousness, but demonstrates its application in areas other than physics. We have said that "perceptual" consciousness consists of five separate sensory realms macroscopically coordinated by five corresponding dimensions; but we have not limited consciousness to perception. There may be another realm or realms, other than perception, coordinated with the perceptual realms in the same manner as they are coordinated amongst themselves. An "observational realm" may exist that not only accounts for the existence of observers, but represents "potential perception" in some manner, and thus corresponds to a dimension the way that the sensory realms correspond to dimensions. If so, it should account for why you move and talk in a way that suggests consciousness "in" you, and why I see you in space and in time. Its dimension should be interchangeable with space, time, and mass. Above all, it should show why you do not need a solipsism to demonstrate the unity of consciousness, that there is a "you *and* I" as well as a "you or I." It should show that there is no need to identify consciousness with yourself or myself. This is a tall order; if it can be accomplished without new metaphysical assumptions, it should be taken seriously.

The Observational Realm

To begin with, we will have to show that there is a distinct *physical* difference between observation and perception. Perception is what I see myself, while observation is what I hear you telling me about what you see. (You may be the "I," and I the "you," from your standpoint.) Socially, morally, culturally, these two are equal; we have been trained since early childhood to accept the experience of others as equivalent to our own. But physically, they are different. What I see consists of photons, while what you are telling me you see consists of words. Perceptions and observations are entirely different physical phenomena, even if they concern the same object.

My experience and yours become *physically* equivalent only when I translate my perceptual experience into observational experience, through the use of words or other symbols. Perception is experienced in the sensory realms, while observation is experienced in its own separate realm, even though the *information* of which it consists is experienced in the sensory realms. In other words, I hear what you are telling me through the auditory realm, or see what you have written in the visual realm, but what I am experiencing is more than sound or light. I am experiencing your experience through a *medium* based in sensory perception. Observational experience is something entirely distinct from perception, but it is at the same time *reducible* to perception. All I am actually *perceiving* are visions and sounds, even though I am experiencing much more. The sensory reduction of observational experience to sound or light (or even touch) is similar to the tactile reduction of sound and light themselves. This has important consequences for the manner in which the observational realm, and observers, are experienced. There are structural analogies between the tactile and "higher" sensory realms on the one hand, and perception and observation on the other. We will explore this later in the chapter.

Observational experience is coordinated with perceptual experience in the same way that the perceptual realms are coordinated with each other: through dimensional potentials. In perceptual consciousness, actual experience in any realm is accompanied by potential perception in every other realm. Where I see an object is where I may hear or touch it. Observational consciousness works the same way; your description of an object's location is exactly the same as my potential perception of the same object. Even if I am in the next room and you are looking at the object and describing it to me, I can perceive it (whether I actually do or not) at the location you describe. There is, therefore, a dimensional coordination of observational with the whole of perceptual consciousness. Observation is *potential perception*. The dimensional coordination of observation with perception is a manifestation of the same structure of consciousness as we have already found among the five sensory realms.

This explains why the concept of matter tends to be reinforced by observational experience. In the same way that matter is assumed to cause seeing, hearing, and touching, matter simultaneously "causes" you and me to see the same thing at the same place and time. Matter serves to explain potential perception among observers in the same way that it explains it among the senses. But, as we have shown matter to be a limited and unnecessary assumption for perceptual experience, it is likewise unnecessary for observational experience.

In the process of eliminating matter we have discovered a structure of consciousness that explains both perception and observation without having to introduce separate metaphysical assumptions for each. In fact, we are able to do away with the troublesome assumption that consciousness has to exist inside of observers. If observation can be shown to be another realm of consciousness similar in structure to the perceptual realms, it does not need to exist inside of me or of you or of anything else. It just is. Observers are manifest in space-time not because they have consciousness "in" them, but because observational information is reducible to sensory experience, of which space-time-mass is the context. What is *perceived* in the form of observers in space-time-mass is the projection of a *six*-dimensional pattern onto five dimensions. The sixth dimension (that which corresponds to the observational realm) appears foreshortened in five dimensions the way the fifth dimension appears foreshortened in four.

A scientific fact is a phenomenon that may be perceived by anyone at any time under the same conditions. Science is, therefore, potential perception, or what I have called "observational consciousness." What I see is not science until I *say* that I perceive it, and you say that you perceive it, and we agree that any other observer from the same perspective will communicate that he or she perceives it. This is a more or less traditional definition of science. But not all information is scientific or "six-dimensional," only that which represents potential perception. If I say "I am thinking of Christmas," or "I am hungry," it is not something that you or anyone else may potentially perceive; thinking or being hungry may be actual experiences but they are not observational experience, and therefore not science. Scientific or observational experience is always dimensionally structured; experience in any one realm is always confirmable by experience in every other realm. It is this structural interconnection of sensory and observational potentials that gives the physical world a sense of greater "reality" than nondimensional, or purely conceptual, experience.

The formal scientific method is, of course, a fairly recent development in the evolution of human consciousness. More rudimentary forms of observational experience go back to the first use of symbols and spoken words. But these were bound by culture and language, and thus potentially perceivable only within a limited context. An "observer" had to know the language and live in the culture to know the meaning of the symbols. Science is (ostensibly, at least) universal observational consciousness in that all humans are allowed to be observers. Science carefully screens and sifts observational patterns from other forms of consciousness and interrelates them on a global scale. That which science says is true thus becomes potential perception at any time and place, in any language or culture.

But in what sense does science consist of "six-dimensional" patterns? Einstein noticed in 1905 that time was an infinite potential, like space, and that it served to coordinate experience in the same way. He established the fundamental identity of space and time in order to create the continuum of "space-time" that he used to explain relativity theory. Time was fairly easy to envision as a dimension because one could picture being "in" time more or less the same way as "in" space. Special relativity was limited to a discussion of the effects of *uniform motion*, or constant velocity, which could be fully explained in terms of the four dimensions of space-time; accelerating reference frames, or *nonuniform motions* were specifically excluded. In the general theory of relativity, however, Einstein went one step further and spoke of gravitation, (which he stated was physically equivalent to constant acceleration), as a "curvature" of these same four dimensions. He did not specifically identify gravity and constant acceleration with a fifth dimension, but an additional dimensional unit is necessary to describe it, as we have noted earlier. If *nonuniform acceleration* is to uniform acceleration what nonconstant velocity is to constant velocity, it must indicate an additional dimension. It is this dimension that corresponds to the observational realm.

Random nonuniform acceleration is the *potential* for observational experience. The increase of entropy throughout the universe with the passage of time as implied in the second law of thermodynamics provides the background for experiencing observational information. *Orderly* non-uniform acceleration is *actual* observational experience. Orderly motion and communication from observers is discernible only because there is a tendency toward disorder everywhere else.

Actual observational experience is science; science, therefore, consists of patterns in six dimensions. As we will see, it is the perspective of the observer himself, as he describes five dimensional patterns that he perceives, that is the dimensional factor added to space-time-mass.

But if nonuniform acceleration really indicates a dimension, how do we measure it? We can measure time, space, and mass; how about measuring "order?" What units shall we use? Here we may have run into trouble; nonuniform acceleration is by nature indeterminate. It cannot be measured, and an observer's behavior within it cannot be predicted. You can measure an observer's length, width, velocity, acceleration, and mass, but you cannot say how he, she, or it, will change acceleration. You can measure the energy he uses to accelerate, but you cannot determine the "curvature" he gives it. (The "curvature" of energy may be either in the form of "work" or of "information.") Measurement is limited to the five perceptual dimensions. Even within these, complete measurement is limited to the macroscopic level, according to the Heisenberg uncertainty principle. Is there a connection, then, between quantum indeterminacy in five dimensions and macroscopic indeterminacy in six? Quantum indeterminacy, we have seen, is due to the tactile reduction of light. Might not macroscopic indeterminacy be due to the sensory reduction of observational experience? On the quantum level, the perceptual realms and their corresponding dimensions are indistinguishable; on the macroscopic level, words are experienced both as sound and symbol, or in two separate realms at once. Therefore the dimensions corresponding to each are neither qualitatively distinct nor quantitatively measurable. The indeterminate behavior of macroscopic observers is in many ways similar to that of subatomic particles. Each reveals the dimensional structure of consciousness at a different level.

Another similarity between subatomic particles and observers is that the behavior of each becomes increasingly determinate in the aggregate. We can know very little about where an electron will go from one moment to the next, but if we have several trillion of them in a copper wire, we can predict with great accuracy how many will move how far. We cannot know which ones will move and which not, only their average behavior. Similarly, we can say next to nothing about where one man on the street might be going, but we can predict with considerable accuracy how many people will go to the druggist or to the pet store on a given day. We have no way of knowing who they will be, but they will show up without fail, nevertheless. Marketing analysts bet good money on it every day. As with subatomic particles, the greater the number of observers, the higher the predictability of aggregate behavior.

A major difference between subatomic particles and observers is that all observers are different while all electrons, protons, etc. are identical. You can tell one man on the street from the next by his looks, his dress, and how he walks, but you cannot tell one pi meson from another; they all "look" and "tend to act" alike. But in making this statement, we should realize where we are looking *from*. We are looking at both the man on the street and the pi meson from the standpoint of macroscopic sensory experience. All pi mesons may look alike due to the structural limitations of the space, time, and mass dimensions with which we attempt to view and measure them, that is, due to the structural limitations of the quantum screen. If we were to view them from the standpoint of their fellows (don't ask me how), rather than "from above," they might each display a unique personality. (There is certainly room for inherent uniqueness within the uncertainty relation.) If we view observers "from above," or from the standpoint of observational consciousness rather than from perceptual consciousness, *they* would all look alike! We would know nothing of their faces or clothing, only what they said they were perceiving. If they were good observers, they would say they were all seeing the same thing from different perspectives. (We would have to dismiss the liars, bigots, and charlatans, of course, but that is just what science is supposed to do.) Real observers, as observers, are all identical, even if some of them wear pink socks and some green.

From the standpoint of observational consciousness, we can even predict the behavior of *individual* observers in terms of probability relations the same way we do for a subatomic particle from the standpoint of perceptual consciousness. We cannot say for sure what a certain man on the street will do today, but our marketing people can tell us how many people buy red ties in New York City on a given day, and as there are 7,628,406 consumers in that market, the particular man we are concerned with has one chance in 5334.55 of buying a red tie today. That is his "wave function."

Interchange of the Observational Dimension

A test for the dimensionality of the observational realm and its coordination with the perceptual realms is dimensional interchange. Let us see if we can interchange it with any or all of the other dimensions. You (or any other good observer) stand in one room while I stand in another room down the hall where I cannot see you. You shout through the wall telling me about the furniture in your room: what you see, hear, touch, and smell and where each item is. All I actually perceive is the sound of your voice, but I experience the whole room, furniture and all, in the observational realm. I have never been there and do not remember it (perceive it at a time value in the past). If my experience is "potential perception," according to our definition, and the dimension corresponding to the observational realm is interchangeable with the other dimensions, I should be able to rotate the axes of perceptual consciousness and perceive directly what you say you are perceiving. I feel "g" forces in my body as I accelerate through the doorway, down the hall, and into your room. Looking around the room, I perceive directly everything you described, where you described it. I have not only interchanged dimensions, but interchanged the right ones, in an orderly manner. I changed my acceleration at just the right times and directions to actualize potential perception.

If I had shouted through the wall instead, describing what I see to you, I would have seen and heard you, as a living observer, move in an orderly way into my room. You would then have observed all the objects in my room where I perceived them.

The interchange described here is a clear definition of the limits of the observational realm as opposed to non dimensional, or conceptual, experience. Observation is often confused with conceptual experience as both are non perceptual and supposed to exist "in the mind." It may appear that an observer's description of an object is a thought as it enters consciousness, but it is a dimensionally-structured thought to the extent that the perceiver can go see it where it is observed. You or I cannot interchange dimensions in the manner described here and see an object that we are thinking about. If observational consciousness is a form of thought, it is, like perception, a dimensionally-structured thought.[22]

A Physical Definition of Life

In seeking a physical definition of life, it is important to realize that we are looking at life from the outside, or how it looks on the "quantum screen," within perceptual consciousness. We are not looking at consciousness itself, but at objects that appear "to be conscious." Nonetheless, I believe that a physical definition of life is possible. Some objects are alive, others are not; all we are doing here is trying to determine the difference in physical terms.

The second time dimension is manifest in changes in velocity, or in the acceleration of massive objects in space-time. Certain objects, however, display not only changes in velocity, but orderly changes in acceleration as well, or *orderly nonuniform acceleration*. Such objects, it will be noticed, are *living observers*; the same objects that are capable of changing the direction or magnitude of their acceleration are also capable of reacting to objects around them, and may communicate observational information from their perspective in space and time.[23] It is only observers, or living objects, that display behavior that tends toward "order," as opposed to entropy.

Entropy is the background of the orderly behavior of observers. It is the context within which information from observers has meaning. The disorderly, or random, behavior of non living objects is the potential within which observational consciousness exists; orderly nonuniform acceleration is actual observational consciousness.

Uniform acceleration is a change in velocity, such as that experienced by a car while you are stepping on the gas (or the brakes), or by a satellite in orbit. The car experiences a linear acceleration, or change in the magnitude of its velocity, while the satellite experiences a centripetal acceleration, or change in the direction of its velocity. If you put a brick on the accelerator, the car will keep gaining velocity at a constant rate (disregarding friction), and if you provide a constant gravitational field for the satellite, it will continue to orbit indefinitely at a constant rate. The acceleration of each will change only when it experiences a change in energy: the car hits a bump, or the satellite runs into air resistance or is pulled by some other gravitational field. In each case the change in acceleration is *random*: the car will crash, or the satellite drift off aimlessly. But if the source of the new energy is an observer, "order" will be imposed on the system, from the standpoint of the observer. The observer will steer the car on the road, or fire retrorockets to keep the satellite in orbit. Only observers are capable of orderly nonuniform acceleration; energy is random unless it is associated with a living organism. Life can be distinguished from nonlife, therefore, not by motion alone, but by a particular type of motion. Life is orderly energy.

This is a physical definition of life. It provides a new perspective on the old problem of understanding life through science; life is within consciousness, and not consciousness in life. But its limitation is that there is no way to quantify or scientifically define "order." Everyone knows that a brick building is more orderly than a pile of bricks, or that obeying traffic signals is more orderly than not obeying them, but there is no way to put a finger on exactly what order is. We cannot measure it, and hence cannot in any way quantify life, the way we can mass, length, or time. If nonuniform acceleration is a dimensional potential, it is not quite the same as the other five.[24]

But the order that exists within potential randomness is always associated with an observer, or observers, in space-time-mass. Order always has a perspective; what is positive order to one observer may be negative to another. One person likes the furniture arrangement, another does not. The termites forming colonies in the beams under your living room floor are creating what to them is a distinctly orderly arrangement of things; you will disagree. Nonliving objects, on the other hand, though they exist in five dimensions and must obey the laws of physics, do not change their acceleration in response to any perceived order in the objects around them.

The order created by observers is local only, and should not be understood as a defeat of entropy as a whole. All living beings require a constant input of new energy from outside sources, the generation of which requires the creation of new entropy. Life on Earth, however orderly, depends on the fusion of hydrogen atoms in the sun, a fuel-burning process that will eventually exhaust itself and contribute to the long-term winding down of the cosmos. The second law of thermodynamics will always hold true for the universe as a whole: disorder will increase with time, despite local pockets of reversal in the vicinity of observers. This would have to be true in any case, as any actual order in the universe can be no larger in terms of space, mass or time, than the potential for order (entropy) against which it is perceived. It should be kept in mind also that the eventual demise of the universe may be considered something of a dimensional extreme, as we still have some billions or trillions of years before it arrives. We may well expect some new anomalies to show up by then.

The limitations of a physical definition of life are in physics, not in life. We have already seen that the dimensional structure of consciousness is limited to macroscopic dimensions. It works well in everyday life, in the middle latitudes of space-time-mass. Reality also exists at dimensional extremes, but appears on the screen only in distorted form. Similarly, the quantum screen presents only a limited view of life. The dimension of life is not a space dimension, and appears on the screen only in foreshortened form, the way a third space dimension appears on a flat surface, or motion appears on a still picture. The identity of *non* living objects is easy to establish through perception because three dimensions appear at the same time. (Only mass and time must be established by more than one space-time perception.) The identity of living objects, however, requires a more complex and tenuous association of informational patterns.

Observational consciousness, or the life that appears to be "in" observers, is a separate and distinct realm of consciousness and cannot ultimately be understood *within* perceptual consciousness. It appears on the quantum screen only in "flattened" form. Trying to see or hear life is very much like a single cell trying to "feel" sound or light. Life in observers can be experienced only when experience is not limited to perception.

Evolution of Observational Consciousness

We are aware of observational information through the senses, but how do we add yet another dimension to the quantum screen?

Observation encompasses all perceptual realms in potential form. At every point on the screen where there is an observer there is a "potential screen." When an observer says he sees an object, *he* is at the point on the quantum screen where we see him and hear the sound of his voice. When another observer tells us about the same object from his perspective, he is on the screen also. Each observer (reports that he) sees the same thing; the only difference between any two observers is *perspective*. Observational consciousness arises when the location of the object is revealed to be the same for each observer once his perspective is factored out. You or I can, of course, see for ourselves if the object is there in actual perception (itself on the screen where the observers say it is), but we rarely do. For the most part, we rely on the honesty of observers, especially if there are a large number of them. It is difficult to lie in the aggregate.

Perspective is essential to the meaning of observational information and constitutes the dimensional factor between perception and observation. The same physical object is observed from any point in space-time-mass but appears different from different perspectives. When perspective is factored out, an additional dimension is imposed on perception and it becomes observation. It is only with this additional dimension that the object is the same for all observers, and only by this means that five-dimensional objects become science.

But what is the relationship between observational consciousness and the quantum screen, our model for perceptual consciousness? Every observer appears on the quantum screen as a point (or range of points), much as do other physical objects. But where physical objects are five-dimensional, observers are six-dimensional; they are capable of orderly nonuniform acceleration in the form of information or motion. They move on the screen as if they perceived objects around them, and we consider them to be "alive." They are alive, but they do not have actual perceptual consciousness "in" them—that is the screen itself, and there is only one screen. As an observer describes what he perceives, he becomes a "potential screen." He is himself perceived directly on the actual quantum screen, but what he tells us becomes a potential quantum screen at the point he occupies in space-time-mass.

A case could be made for a sort of six-dimensional super screen based on an extrapolation of the five-dimensional quantum screen analogous to the quantum screen's own basis on an extrapolation of the four-dimensional photon screen. All of observational experience—all of scientific knowledge—could be said to exist at locations on such a screen. But I think this stretches the screen model too far. Where objects directly perceived in space are experienced much the way they are seen on a television or computer screen, objects and experiences described by others are not so easily experienced in the same manner. We tend to "think" of places and events that we are told about, even if we know they are (or were) potentially perceptible. We do not picture them as clearly. There is no model for observational consciousness as powerful as that for perceptual consciousness because observational consciousness is not as powerful a part of our lives. This, however, is changing. There may well come a time when collective experience is more important than individual experience.

A question remains. If observation corresponds to an additional dimension, why is it that some physical phenomena manifest this dimension and others do not? Why do six-dimensional observers exist alongside five-dimensional objects in the same physical world?

As I suggested earlier, the application of the structure of light to other realms of perceptual consciousness is most likely due to the fact that vision developed relatively late in evolutionary terms *and* because "bits" of visual information (photons) are much smaller than the bits of which other sensory information is composed. We can, therefore, understand hearing *in terms* of potential photons, rather than the other way around. But what if both of these conditions did not apply? What if a new realm of consciousness were to develop whose "bits" were larger than those of existing realms? As the new bits would not be able to constitute a new universal medium, they would have to be projected, in a higher dimensional form, onto the old universal medium. This, I believe, is what is happening with the observational realm. Bits of observational information are the letters and sounds and symbols by which observers communicate; they are all considerably larger than photons. As an example, the tiny lights on a television screen are each composed of many millions of photons. We experience observational information, therefore, as foreshortened six-dimensional forms projected onto the quantum screen along with five-dimensional perceptual objects.

As human civilization evolves, direct perceptual experience becomes less a part of consciousness as a whole. We learn more about the world through other people, and observational experience becomes increasingly important. We talk, read books, watch movies, and gain information through electronic technology. We use the telephone, listen to radio, and watch the world on television. Observers become more important than objects. Experience through others becomes less distant; potential perception becomes a form of actual experience. You or I do not have to be there to experience what is happening in the world. We experience Asia, Africa, and the far side of the moon without ever perceiving it. Individual experience becomes more and more absorbed in collective experience: the perspective of *actual* perception is factored out along with that of potential perception. The quantum screen appears more and more like a point on a six-dimensional matrix. We each see ourselves individually as small parts of a greater collective whole.

Media such as books, magazines, newspapers, telephones, television, and radio promote the evolution of observational consciousness. Particularly interesting among these is television. I am thinking here not of television programming, but of television as a medium. What we experience on a television screen is what we would see if our eyes were where the camera is. But television is different from other observational media in that it utilizes light, the universal medium of perceptual consciousness. This means that we experience directly what is happening on the screen; there is no need to "envision" a situation on the quantum screen, as is necessary with the print or radio media. Points of light constitute images on a television screen in the same way that photons constitute images on the retina; observation becomes perception.

Every other communications medium requires a conceptual operation for factoring out perspective; we have to think of a human writer or radio announcer in a space-time setting in order to appreciate what he or she observes. But television images need not be factored at all. Television information is not processed by an observer: in fact, there is no observer. Observation is experienced directly; everyone watching a live television broadcast experiences the same images, from the same perspective, at the same time. There is among televiewers, therefore, not only common experience, but common perspective. The advent of television brings more "wholeness" to observational consciousness than any other medium.

The continuing evolution of observational consciousness is the greatest single fact of our time.

VIII
CONCLUSIONS

Scientists are so close to the work they are doing that they rarely appreciate the immense picture they are unveiling little by little. Most remain comfortable with metaphysical assumptions that keep their work going. If they question them at all, they do not question them often. An independently existing material world remains strong among them, however cracked and contorted by new discoveries, because it has for centuries provided them with a durable metaphysical foundation upon which to build their world. To undermine that foundation now may seem meddlesome and potentially destructive to the work going on above.

But I do not mean to undermine their work. I hope merely to redo a foundation that has been undermined already and can no longer withstand the weight above it. This can be done, for the most part, without damage to the superstructure. Those working above may feel an occasional tremor, but they need not be disturbed overall. Their immediate surroundings will change very little. They may look up one day, though, and find that the entire building no longer stands where it once stood.

Can I call the work I present here science? Is the foundation part of the building? The simple answer is yes; the building cannot exist without a good foundation. But the foundation is not where people live and work and have their being. It lies well beneath the conscious level; it is there to provide an interface between the dark, uncertain forces of rock and soil below, and the rational, rectilinear lines and angles of the building above. Those in the building spend little time thinking about the foundation, and that is as it should be. The great power of science is that it need not continually justify fundamentals: "F" always equals "m" times "a" and always will, and we already know what force and acceleration are. The work of one scientist can thereby build dependably on that of others; each need not define terms every time he says something.

If there are people standing on the ground wondering what force and acceleration "really" are, they do not impede progress. Science uses operational definitions to keep the job going, even when nobody knows what it is that is going. The building, if well constructed, will span small gaps in the foundation. It is only with fear of general collapse that workers look down and wonder just what it is that holds them up. If it is new foundation work they need, they will say yes, that it is science, go ahead with it.

But to work on the foundation, we must dig away the soil and rock that lies in the way, and examine *it*, too. We must know more about the ground in order to support the building. Some scientists will say no, that is not part of the building. The work we do and the materials we use are the building; rock and soil are something else. Keep it away from us.

I cannot insist, therefore, that the idea presented here is science. It is an interface between science and that part of life that is not science. It behaves like a scientific theory in some ways, but not in others. It is logical, and its scope is wide: it explains a number of diverse physical phenomena that have not been encompassed before by a single, consistent concept. But I am not sure that it is provable, nor that it will lead to new research in the traditional sense. I hope that predictions based on it will one day be proven or disproven, but this may not happen. Scientists will eventually agree that it is right, but they may never be able to "touch" it, the way they can other theories. The scientist working high on the superstructure may never be able to look down and understand the Earth beneath him in terms of the bricks and two-by-fours that he is used to; all I can hope is that he will know where they come from.

He may say that philosophy never gets off the ground. I will agree and point out that his building *is* the ground, in specially structured form. He will say that my insistence that the building does not exist is absurd. I will say that it exists only because he made it. He will say that by lumping science in with thought, imagination, and opinion, I have ruined its unique properties, as if mixing jewels and dirt. I will say that jewels are dirt. His interest is in what he can see and be sure of, in what will get him through the day; he understands life in terms of science. My interest is understanding science in terms of life.

Western civilization has built an expanding universe within the human mind. The world of science is human imagination, carefully contained in an explosion of space and time, balanced by mass. Science creates what it discovers. At a small corner and to the side is the life process, a cross section of life itself.

The great weakness of science is that it is contained at all and cannot see itself as such. It cannot go outside without getting cold.

ENDNOTES

Images and Realms

1. "Behaviorists" would deny this, claiming that thought, feelings, and consciousness itself are reducible to reflexes and thereby to analytical consideration. This is reductionism at its most absurd.

Potentials

2. Signh, Jagjit, *Great Ideas in Information Theory, Language, and Cybernetics* (New York: Dover Publications, 1966), 12.

3. Einstein, Albert, *Out of My Later Years* (New York: The Philosophical Library, 1950). I should apologize to Mr. Einstein for using his thoughts and observations for my own purposes, especially since my conclusions differ from his, but I do not think he would mind. I believe he is speaking in this passage of what he would consider the "mental side" of physical reality that corresponds to the "material side." He would say that matter exists independently of actual experience; I differ from him only in interpreting matter as a form of potential experience.

In the same work, Einstein ultimately despairs of understanding the conceptual context of sensory perception:

> The very fact that the totality of our sense experiences is such that by means of thinking (operations with concepts, and the creation and use of definite functional relations between them, and the coordination of sense experiences to these concepts) it can be put in order, this fact is one which leaves us in awe, but which we shall never understand . . .
>
> In my opinion, nothing can be said concerning the manner in which the concepts are to be made and connected, and how we are to coordinate them to the experiences.

The Macroscopic World

4. A case could be made for the existence of an observational realm among animals, in that they communicate with one another. But it is doubtful that they are able to communicate observations in a dimensional context to nearly the same extent as people. An animal cannot describe what he is

looking at well enough to create among his fellows what we will call "potential perception."

5. Barrington, E.J.W., *Invertebrate Structure and Function*, 2nd ed. (New York: John Wiley and Sons), 388. We could go back even before the tactile realm (before there were cell membranes), and imagine we are complex amino acids "tasting" our way through the primordial soup.

Barrington seems to suggest that tactile perception is reducible to chemical stimuli: "An active control of potassium concentration is a widespread feature of living systems, and it is likely . . . that polarization of the surface membrane is a common, and perhaps universal property of living matter. This suggests that from an early stage of evolution the primary effect of environmental disturbance was to create localized states of instability in surface membranes, involving changes in their ionic permeability. This would have resulted in a flow of ions which could bring about some measure of depolarization."

Touch may be, therefore, "no more than" an elaborate form of taste. But it constitutes a separate realm of perception in that it consists of a distinct form of information in a dimensional context.

6. Brown, Evan L. and Deffenbacher, Kenneth, *Perception and the Senses* (New York: Oxford University Press, 1979), 221. Brown and Deffenbacher have shown in hearing tests that the perception of the third space dimension depends on the establishment of object identity in previous perceptions: "We are much less accurate at determining the distance of sound source than its direction. In addition . . . we must be familiar with the sound source."

7. I do not mean by this that time is constructed as a potential for chemical information the way space and mass dimensions are constructed from time.

8. Burtt, E.A., *The Metaphysical Foundations of Modern Science* (Garden City, NY: Doubleday and Co., 1954), 83–84. According to Galileo, the "primary qualities" of perception are number, figure, magnitude, position, and motion. He describes these as absolute, objective, immutable, and mathematical.

9. Pais, Abraham, *Subtle is the Lord* (New York: Oxford University Press, 1982), 329.

10. Burtt, 109. Burtt quotes this passage from *The Philosophical Works of Descartes*, trans., Haldane and Ross (Cambridge, 1911), I:61.

11. Except to the extent that we have improved upon the Mach Principle.

The Enigmas of Modern Physics

12. By the late 1800's there was very little evidence for Newton's "corpuscular" theory of light. Research to that date strongly suggested that light consisted of waves.

13. I do not mean a physical medium for the *transmission* of light, such as the ether, but light itself as a carrier of information. Marshall McLuhan stated that as a medium, a light bulb is one hundred percent information. This is what I mean.

14. Planck discovered quanta, but the credit for discovering the "quantum effect" of light, and the formula E = hυ, belongs to Einstein.

15. Not all physicists agree that space is "quantized" in the way I am suggesting here. But I would contend that we can close the gaps between quanta in space only by opening them between time or mass.

16. Planck never said anything about a "screen." I am using the "quantum screen" as a conceptual tool to express in a different way his discovery that energy is "quantized."

17. Pais, Abraham, *Subtle is the Lord* (New York: Oxford University Press, 1982), 180. Einstein demonstrated in his theory of general relativity the "complete physical equivalence of a gravitational field and the corresponding acceleration of the reference frame."

Light

18. How do we justify a "mass" value for a photon? Normally, photons are understood to have momentum, but no mass, as they must accelerate instantaneously in space-time. (They can travel at no intermediate velocity between 0 and "c.") But here we do not put photons in space-time, and may thereby assign them a mass value. In fact, we *must* assign them a mass value if we are to understand them as tactile sensations. We can do this with a bit of the same mathematical trickery that is used to assign them momentum.

The momentum of an object is equal to its mass multiplied by its velocity. The momentum of photons, however, is a "derived" quantity in that it can be calculated, but not measured directly. No one has ever caught a photon, weighed it, and multiplied this value by its velocity. Instead, a photon's momentum is determined by measuring the momentum of a particle with which it collides. By the law of the conservation of momentum, the photon is assigned the same momentum as that measured in the particle. The momentum of a photon has a physical meaning, but it is not directly observable.

According to Einstein's formula "E = hυ," we can calculate a photon's energy by knowing its frequency. By dividing energy by velocity, we get the photon's momentum as "hυ/c," or its energy divided by its "velocity." (Energy is equal to mass times velocity *squared*, so momentum is equal to energy divided by velocity.)

We can derive the "mass" of a photon by taking this one step further. If the momentum is divided again by "velocity," the "mass" of a photon may be expressed as:

$$m = h\upsilon/c^2$$

where "m" is the photon's mass, "υ" its frequency, "h" the Planck constant, and "c" the "velocity" of light. A photon's mass depends on its frequency.

The reader may notice that we are dividing momentum by a "velocity" that we have said is not a velocity at all. This is exactly the point: there can be no velocity without space and time. If "c" is the relation between space and time, it is space and time that are *created* as photon mass becomes energy. The universe we know comes into being as minute tactile sensations are arranged into information in a dimensional context.

19. I believe that a space-time world based on sound could be demonstrated mathematically and perhaps demonstrated experimentally. It would consist of two, rather than three, space dimensions.

20. I am speaking here of energy quanta, not protons or electrons or other subatomic particles sometimes referred to as "quantum" particles.

21. The situation is complicated by the fact that, as was mentioned previously, the body is never experienced as an indivisible whole nor as a distinct part. We are always accelerated somewhat when we touch something, and we must be touching something whenever we accelerate.

Observational Consciousness

22. I do not rule out the existence of other realms, dimensions, and interchanges. Another type of structural relation is that between conceptual and observational consciousness; an object may be conceived and then created in observable form.

23. Animals, as they are capable of orderly nonuniform acceleration, are observers according to this definition. But symbolic communication, as we normally understand it, does not exist between humans and animals. Accordingly, the special structure of observational consciousness we call science must exclude animal observers. The existence of animals in the same dimensional context as humans can be interpreted to mean that communication remains potential in the nonsymbolic, nonscientific sense, especially among animals themselves.

24. Order *can* be defined in terms of probability. The less probable an arrangement, the more orderly. We may be in a situation in which we have identified a potential, but not yet fully dimensionalized it; we know the probabilities, but have not yet developed it into a pattern in a dimensional context.

About the Author

Samuel Avery lives with his family on a farm in Hart County, Kentucky, along the banks of the Nolin River. He spends his time gardening, building, and teaching.

Printed in the United States
33388LVS00003B/33